Ecological Exile

Ecological Exile explores how contemporary literature, film, and media culture confront ecological crises through perspectives of spatial justice – a facet of social justice that looks at unjust circumstances as a phenomenon of space. Growing instances of flooding, population displacement, and pollution suggest an urgent need to re-examine the ways social and geographical spaces are perceived and valued in the twentieth and twenty-first centuries. Maintaining that ecological crises are largely socially produced, Derek Gladwin considers how British and Irish literary and visual texts by Ian McEwan, Sarah Gavron, Eavan Boland, John McGrath, and China Miéville, among others, respond to and confront various spatial injustices resulting from fossil fuel production and the effects of climate change.

This ambitious book offers a new spatial perspective in the environmental humanities by focusing on what the philosopher Glenn Albrecht has termed 'solastalgia' – a feeling of homesickness caused by environmental damage. The result of solastalgia is that people feel paradoxically ecologically exiled in the places they continue to live because of destructive environmental changes. Gladwin skilfully traces spatially produced instances of ecological injustice that literally and imaginatively abolish people's sense of place (or place-home). By looking at two of the most pressing social and environmental concerns – oil and climate – *Ecological Exile* shows how literary and visual texts have documented spatially unjust effects of solastalgia.

This interdisciplinary book will appeal to students, scholars, and professionals studying literary, film, and media texts that draw on environment and sustainability, cultural geography, energy cultures, climate change, and social justice.

Derek Gladwin is a SSHRC Postdoctoral Fellow at the University of British Columbia, Canada. He was an Environmental Humanities Fellow at the University of Edinburgh, UK, and at Trinity College Dublin, Ireland. His authored and co-edited books include *Contentious Terrains* (2016), *Unfolding Irish Landscapes* (2016), and *Eco-Joyce* (2014). Please visit derekgladwin.com.

Routledge Environmental Humanities

Series editors: Iain McCalman and Libby Robin

Editorial Board
Christina Alt, St Andrews University, UK
Alison Bashford, University of Cambridge, UK
Peter Coates, University of Bristol, UK
Thom van Dooren, University of New South Wales, Australia
Georgina Endfield, University of Nottingham, UK
Jodi Frawley, University of Sydney, Australia
Andrea Gaynor, The University of Western Australia, Australia
Tom Lynch, University of Nebraska, Lincoln, USA
Jennifer Newell, American Museum of Natural History, New York, US
Simon Pooley, Imperial College London, UK
Sandra Swart, Stellenbosch University, South Africa
Ann Waltner, University of Minnesota, US
Paul Warde, University of East Anglia, UK
Jessica Weir, University of Western Sydney, Australia

International Advisory Board
William Beinart, University of Oxford, UK
Sarah Buie, Clark University, USA
Jane Carruthers, University of South Africa, Pretoria, South Africa
Dipesh Chakrabarty, University of Chicago, USA
Paul Holm, Trinity College, Dublin, Republic of Ireland
Shen Hou, Renmin University of China, Beijing, China
Rob Nixon, Princeton University, Princeton NJ, USA
Pauline Phemister, Institute of Advanced Studies in the Humanities, University of Edinburgh, UK
Deborah Bird Rose, University of New South Wales, Sydney, Australia
Sverker Sorlin, KTH Environmental Humanities Laboratory, Royal Institute of Technology, Stockholm, Sweden
Helmuth Trischler, Deutsches Museum, Munich and Co-Director, Rachel Carson Centre, Ludwig-Maxilimilians-Universität, Germany
Mary Evelyn Tucker, Yale University, USA
Kirsten Wehner, National Museum of Australia, Canberra, Australia

The *Routledge Environmental Humanities* series is an original and inspiring venture recognising that today's world agricultural and water crises, ocean pollution and resource depletion, global warming from greenhouse gases, urban sprawl, overpopulation, food insecurity and environmental justice are all *crises of culture*.

The reality of understanding and finding adaptive solutions to our present and future environmental challenges has shifted the epicenter of environmental studies away from an exclusively scientific and technological framework to one that depends on the human-focused disciplines and ideas of the humanities and allied social sciences.

We thus welcome book proposals from all humanities and social sciences disciplines for an inclusive and interdisciplinary series. We favour manuscripts aimed at an international readership and written in a lively and accessible style. The readership comprises scholars and students from the humanities and social sciences and thoughtful readers concerned about the human dimensions of environmental change.

"Derek Gladwin's *Ecological Exile* is a smart intervention that emerges brightly from the 'energy humanities', a fast-rising area of study which is focusing on the power of contemporary literature, film, and media culture to reveal how human societies produce corrosive infrastructures that, currently, tend to deny the contribution of carbon-based energy systems to social, spatial, and ecological injustices. Engagingly, the book illustrates why the arts and humanities are crucial to scientific and technical debates surrounding fossil fuel production and to stimulating imagination of human behaviors and activities that might be transformed to ensure a thriving, just and survivable future."

—Joni Adamson, Professor of Environmental Humanities, Department of English, and Director, Environmental Humanities Initiative, Arizona State University, USA

"In *Ecological Exile*, Derek Gladwin uses a dynamic, transdisciplinary approach in bringing together the environmental humanities, spatial justice, and cultural studies more broadly. Gladwin shows how modern literary and visual texts respond to ecological crises, altering the ways we imagine ourselves and our environment, and helping us map the shifting terrain of our world system. The result is a significant contribution to contemporary cultural theory and environmental criticism."

—Robert T. Tally Jr, Professor of English at Texas State University, USA, author of Spatiality

Ecological Exile

Spatial Injustice and Environmental Humanities

Derek Gladwin

First published 2018
by Routledge

2 Park Square, Milton Park, Abingdon, Oxfordshire OX14 4RN
52 Vanderbilt Avenue, New York, NY 10017

Routledge is an imprint of the Taylor & Francis Group, an informa business

First issued in paperback 2019

© 2018 Derek Gladwin

The right of Derek Gladwin to be identified as author of this work has been asserted by him in accordance with sections 77 and 78 of the Copyright, Designs and Patents Act 1988.

All rights reserved. No part of this book may be reprinted or reproduced or utilised in any form or by any electronic, mechanical, or other means, now known or hereafter invented, including photocopying and recording, or in any information storage or retrieval system, without permission in writing from the publishers.

Trademark notice: Product or corporate names may be trademarks or registered trademarks, and are used only for identification and explanation without intent to infringe.

British Library Cataloguing-in-Publication Data
A catalogue record for this book is available from the British Library

Library of Congress Cataloging-in-Publication Data
Names: Gladwin, Derek, author.
Title: Ecological exile : spatial injustice & environmental humanities / Derek Gladwin.
Description: Abingdon, Oxon ; New York, NY : Routledge is an imprint of the Taylor & Francis Group, an Informa Business, [2018] | Series: Routledge environmental humanities | Includes bibliographical references and index.
Identifiers: LCCN 2017027086| ISBN 9781138189683 (hbk) | ISBN 9781315641478 (ebk)
Subjects: LCSH: Environmental justice. | Pollution—Social aspects.
Classification: LCC GE220 .G534 2018 | DDC 363.7—dc23
LC record available at https://lccn.loc.gov/2017027086

ISBN: 978-1-138-18968-3 (hbk)
ISBN: 978-0-367-27111-4 (pbk)

Typeset in Times New Roman
by Swales & Willis Ltd, Exeter, Devon, UK

Contents

Figures

Acknowledgements

I am indebted to the following people who have assisted me with constructive feedback in some form or another: Nessa Cronin, Christine Cusick, Danine Farquharson, David Farrier, Graeme Macdonald, Miguel Mota, Maureen O'Connor, Matthew Reznicek, and Robert T. Tally Jr. Further thanks go to Patricia Miranda Barkaskas, who, in addition to showing consummate emotional support, offered significant feedback during the writing and revision of the book. There are many others who I have informally chatted with about this book, but they are too numerous to name. Suffice it to say, thank you! Everyone's encouragement and feedback were essential to the completion of this book. Moving forward, I fully own whatever mistakes or oversights no doubt still remain in the text.

I am also appreciative of the comments from audience members in international conferences, workshops, roundtables, and public talks where some of this work has been presented at the University of British Columbia, University of Glasgow, University of Edinburgh, Edinburgh College of Art, National University of Ireland, Galway, University College Cork, Trinity College Dublin, Memorial University Newfoundland, Simon Fraser University, and Concordia University (Montreal). I am also immensely appreciative for the funding support that came from the Social Sciences and Humanities Research Council of Canada (SSHRC) to work in the Department of English at the University of British Columbia, as well as supplemental research and intellectual support from places I held Visiting Research Fellowships: the Institute for Advanced Studies in the Humanities (IASH) and the Edinburgh Environmental Humanities Network (EEHN) at the University of Edinburgh; the Moore Institute for Research in the Humanities and Social Sciences at NUI, Galway; the School of Irish Studies at Concordia University (Montreal); and the Centre for Environmental Humanities (CEH) at Trinity College Dublin. I additionally want to acknowledge the Ireland Canada Research Council, which generously funded my Visiting Fellowship at Trinity College Dublin, and where I organised a public event on climate change and gave a public lecture.

All of the figures used in this book received permissions for reproduction. I want to graciously acknowledge the Land Art Generator Initiative, Oliver Comerford, Roseanne Watt, Greenpeace UK, Don't Panic London, Brett Bloom, and David Katznelson. See the list of figures for more information on images. Permissions were also kindly granted for the reproduction of the following published poetry and drama: John McGrath, *The Cheviot, the Stag, and the Black, Black Oil*, edited by Graeme Macdonald (London: Bloomsbury, 2015); Eavan Boland, 'In Our Own Country', *Domestic Violence* (Manchester: Carcanet Press, 2007); Gillian Clarke, 'Cantre'r Gwaelod', *Zoology* (Manchester: Carcanet Press, 2017); as well as the unpublished poetry (some of which was on *The Guardian*'s 'Keep it in the ground: a poem a day' website) by Theo Dorgan, Rachael Boast, Jackie Kay, Carol Anne Duffy, and Roseanne Watt, all of which followed fair use guidelines. Short sections of Chapters 2 and 9, although substantially different, emerged from other work published as 'Ecocritical and Geocritical Conjunctions in North Atlantic Environmental Multimedia and Place-Based Poetry', in Robert T. Tally Jr. and Christine Battista (eds), *Ecocriticism and Geocriticism: Overlapping Territories in Environmental and Spatial Literary Studies* (New York: Palgrave Macmillan, 2016) and 'Ecologies of Place in the Irish Short Story', *Canadian Journal of Irish Studies* 40.2 (2017).

Finally, I want to recognise my editors at Routledge, Rebecca Brennan and Leila Walker, for their patience and timely feedback throughout the process, as well as the editors of the Routledge Environmental Humanities series, Iain McCalman and Libby Robin, for their initial interest in the book and its inclusion in the series. The peer reviewers of this manuscript also offered valuable suggestions that helped me expand the parameters of this book and rethink new directions. The copy-editors also provided beneficial revisions on style, sentence structure, and word choice. No worthwhile book is ever published without the time and care given by a competent publishing team. Thank you!

Introduction

Decoding spaces of ecological injustice

This book is highly interdisciplinary – an approach often praised in theory more than in practice. Interdisciplinary scholars assume that integrating knowledge and methods from various disciplines can create a compelling synthesis of ideas to arrive at new findings. This book is also transdisciplinary, creating a conversation between and across a range of critical methods and frameworks.[1] Forging connections between seemingly disparate disciplines and methodologies create some obvious obstacles. These challenges reflect the difficulty of compartmentalising discrete and yet complementary concepts in areas of study such as ecology, social justice, geography, culture, and the humanities, even though they are all, in their own way, transdisciplinary.

Taking such a direction effectively addresses the current environmental crisis, which is ultimately the primary subject of this book, because it mirrors the aim of ecological thinking: examining interconnectedness and interrelationships as part of a whole and unifying existence. This critical path is also beneficial because it draws connections between disciplines rather than maintaining divisions, fashioning unity between divergent fields instead of safeguarding traditional disciplines at the cost of developing further progress on a particular set of issues under examination. My intent, then, is to produce some transdisciplinary conversations between critical fields of study by drawing on various intellectual and social frameworks to confront the planetary environmental crisis. The diverse subject in the following pages of this book dictates its form and agenda.

By way of introduction, I want to acknowledge this inter-/transdisciplinary approach results in some of the following research questions, which, to use a spatial metaphor, has led me down a rabbit hole that has turned into a rather large cavern. What happens when people feel dislocated from the places they live because of environmental damage? How do increasing occurrences of ecological disasters caused by social injustices destroy people's sense of place? Can people feel exiled in the places they live (while never leaving) because of environmental change? How does artistic production in the arts and humanities underscore values and perceptions around personal and collective responses to place amidst ecological degradation? Can a film or a novel, for example, mobilise environmental action by reclaiming narratives in social and cultural spaces?

Addressing questions such as these, this book explores how contemporary British and Irish literature, film, and media culture confront ecological crises through perspectives of spatial justice – a facet of social justice that looks at unjust circumstances as a phenomenon of space. In addition to asking *what creates* environmentally linked spatial injustices, this book explores *who responds* to these unjust circumstances in society, in what cultural forms and geographical areas, and why. In doing so, the purpose is to synergise two interlinking methodologies – the environmental humanities and its relationship to spatial justice – particularly as they both are represented in print and visual works of literature, film, and media.

Reproducing ecospaces

Growing instances of flooding, population displacement, and pollution, for example, suggest an urgent need to re-examine the ways social and geographical spaces are perceived and valued in the twentieth and twenty-first centuries. Although spatial perspectives are vital to analyses of society and culture, they remain less explored in the environmental humanities than in geography or social theory. Regardless, writers, poets, filmmakers, and visual artists are often among the first to anticipate and respond to pertinent social crises in complex and sophisticated creative forms of print and visual texts. These creative forms can socially produce new ways of imagining space in the places we live. They appeal to how society uses narratives, images, and metaphors as a way to frame values and emotions, which have the effect of reproducing social consciousness about environmental issues. Storytelling, for instance, remains crucial for humans to perceive and respond to the world around them. Functioning as a socially conscious criticism, storytelling can serve as a narrative form of resistance. Storytelling is also a spatial practice because it reproduces different ways of seeing or being in the world. As a cultural and feminist theorist, bell hooks understands the power of storytelling as a site of resistance within space: 'Spaces can tell stories and unfold histories. Spaces can be interrupted, appropriated, and transformed through artistic and literary practice' (1990, 152).

A range of creative works in the humanities help mobilise particular diffuse and direct forms of addressing environmental injustices (among other practices of social injustice not covered in this book). Because the humanities are vital to social and political roles in the arts, they have the potential to induce social action by critically engaging with and thinking about the ways societies perceive and value traditions, laws, and public policy. The postmodern geographer Edward Soja speaks to this point when he suggests that:

> we make our geographies just as it has been said that we make our histories, not under conditions of our own choosing but in the material and imagined worlds we collectively have already created – or that have been created for us. (2010, 18)

Changing the process of how people socially construct, deconstruct, and then reconstruct their lives directly impacts unjust social formations of power, whether through voting, organising, or serving one's community. As the acclaimed British filmmaker Pratibha Parmar observed, 'The appropriation and use of space are political acts' (1990, 101). Therefore, my aim in this book is to analyse some of the socio-economic and political practices that both produce and are produced by various forms of spatial relations in specific literary and visual works. If the ecological crisis is largely a result of corrupt social and spatial systems of power, then this has direct bearing on how people may perceive and respond to social and spatial conditions that effect the environments around them.

Maintaining that ecological crises are largely socially produced, I consider how literary and visual texts respond to and confront various spatial and environmental injustices resulting from fossil fuel production and the effects of climate change. This thesis stresses how artistic production contributes to spatial as much as social or historical ontologies, as ways of being in and responding to the world. It emphasises what humans produce also shape and produce them. Social institutions (private and public) that construct unjust spaces not only generate environmental issues – largely because 'nature' has always been socially constructed to support its value as a human product – but are also inseparable from certain kinds of artistic production responding to such injustices as counter-narratives. Various print and visual texts produce their own micro-spaces within social spaces through these narratives to re/create place or home. As a spatial theorist, Henri Lefebvre remained convinced that everyday lives, or life lived as a project, are crucial real and symbolic acts in changing social spaces. The 'body along with space, in space, and as the generator (or producer) of space' is embedded in 'non-formal knowledge' of 'poetry, music, dance and theatre' (Lefebvre 1991, 406–7). To this effect, productions and reproductions of spatial in/justices are often addressed or confronted through artistic forms as social and political action. We can and do make, shape, and construct real and imagined spaces in society through various literary and visual texts.

In *The Political Unconscious* (1981), Fredric Jameson asserts that artistic forms and cultural texts function as 'symbolic acts' that do not exist in political isolation. Rather, the frameworks in which contemporary literary and visual texts reside are equally if not more influential than the text itself. This well-known approach in literary studies, which reinforces that artistic choices traditionally read as aesthetic modes of production and function as historical cultural practices, also enables interpretation through spatial practices and theory. 'Our living', argues hooks, depends upon theorising our experiences 'aesthetically' and 'critically' to produce 'radical cultural practice' (1990, 149). An essential and yet often neglected element of the social and historical dimension of cultural texts is the spatial. Although space may not be a text in the strictest sense, it functions as a real and imagined site that is often dialectically produced without privileging subjective categories of fear and desire, private and public, or built and non-built that are historically documented and socially formed and produced. Textual analysis allows for spaces or places to be read or interpreted as much as print of visual text to be read within these spaces (Smyth 2001, 14–15). Place-based art emerges out of and in response to specific place-based issues or contestations. Artistic production of literary and visual texts become polyvocal, responding to historical traces underlining spatially unjust spaces. Thus, political perspectives are, more generally, 'the absolute horizon of all reading and all interpretation' of artistic forms (Jameson 1981, 17), but viewed as spatially symbolic acts in connection to political ecologies.

Jameson functions here as an effective introductory interlocutor to explain the political dimension of literary and visual texts. His still relevant assertions about politically symbolic acts, despite and even perhaps in spite of his promoting a historicist approach, applies to other political interpretations of artistic forms in the environmental humanities because they accentuate the real and imagined interplay among texts.[2] The ecospatial elements of my approach complement what has been already termed 'ecohistoricist', which, as the name alludes to, scrutinises the historical timeline of climate and ecological injustices related to culture (Trexler and Johns-Putra 2011, 195). Politically symbolic acts apply to spatial dimensions as much as to historicism. Texts are rooted in place and respond to specific places often manipulated by capitalist production, for example, as much as they are framed historically or socially. As a scholar of intercultural anglophone studies, Sylvia Mayer has similarly pointed out how 'texts' can 'direct our perception' because 'they suggest categories of interpretation and evaluation, they function as premises for subject and identity formation, for the creation of ethical systems' (2006, 112). The public and artistic dimension of print and visual texts within the humanities both form identity and create 'ethical systems' because of their social process and production. As the literary and visual texts discussed in this book show, artistic production reclaims the ways texts engage with political acts of environmental and spatial injustices.

Although *Ecological Exile* relies on historical and social formations throughout, its central thesis emphasises how spatial practices magnify political interpretations of texts that exhibit forms of ecological injustices. In other words, the book is primarily an ecospatial interpretation of literary and visual culture. Studying a combination of print (novels, short fiction, poetry, and drama) and visual (film, media, and performance) texts as creative and aesthetic experiences of 'lived life', I examine spatial injustices in three thematic sections: space, oil, and climate. Energy futures and global temperatures, for instance, serve as two of the most significant political and environmental concerns for human and nonhuman survival. As demonstrated by the 2008 global financial crisis, unjust distributions of wealth, resources, and land, coupled with oppression and domination, reached a breaking point for the planet.

Acts of resisting spatial injustices related to environmental and social distresses are not only relegated to politics and legislation; resistance also includes responses within the humanities and the many ways artistic production can permeate the lives of people and influence social change. As Gerry Kearns asserts in the introduction to his timely co-edited volume *Spatial Justice and the Irish Crisis*, 'The crisis is one of ideas as well as of policy' (2014, 3). Ideas of equality and fairness often emerge from the zeitgeist of creative and social change, and in turn are catalysed by people's actions. The imagination, influenced by perception and experience, functions as an active social process. Real and imagined spaces in artistic production, as Lefebvre indicates, can also 'be read or decoded' as much as a geography, idea, or experience (1991, 17). They are interconnected and intertwined.

Contributing to mounting transdisciplinary work in the environmental humanities, I want to add another dimension to the book's aim by focusing on what the philosopher Glenn Albrecht has termed 'solastalgia', or a feeling of homesickness caused by environmental damage in the places people live, and particularly in the ways literature and visual culture have documented the ecological effects of the loss of place. Ecology contains etymological roots pertaining to the concept of 'home' and 'place' (i.e. the study of *oikos* meaning 'house' or 'environment'). Increasing occurrences of ecological disasters literally and imaginatively abolish people's sense of place (or 'place-home'), highlighting why these occurrences are instances of place injustice. Environmentalism as a social movement largely addresses local, regional, and global forms of ecological dispossession and displacement. Ecology interconnects with everything else horizontally, whereas socially produced causes of ecological degradation are bound up by power structures functioning vertically (Soja 2010, 54). Unjust spatial circumstances form unnecessary barriers for people trying to access basic rights. In response, they must shape their own needs on multiple scales (locally, regionally, and globally), specifically when certain forces have exacerbated ecological degradation, and challenge how people might reclaim their own social and physical geographies. The result I want to stress is that people feel paradoxically 'ecologically exiled' in the places they continue to live but are significantly changed because of environmental damage. Instances of solastalgia capture experiences of feeling ecologically exiled, creating eco-anxieties in the places people live, and constitute a major theme in this book explained more in Chapter 2.

There are ways to address and minimise the loss of place-home as a result of spatially induced ecological injustice. If social and ecological circumstances can be altered through the representation and reproduction of space, then spatial injustices can also be overcome through producing places in social narratives of literary and visual culture. Because of the ways space occupies various forces vying for power, and because space is ambivalent, often unfinished and incoherent, it can exist as sites of resistance as well as domination (Smyth 2001, 16). The ways print and visual texts depict, narrate, and value both physical and imaginative spaces influence how people who read or watch such texts identify their existential and creative relationships to environments.

Ecological Exile ultimately addresses the ways literature, film, and media culture in the environmental humanities respond to, confront, affirm, or resist forms of spatial injustice related to dispossession of place, fossil fuels, and climate change. I support the position that the current environmental crisis emerges from socially produced spaces over the past few centuries, relating to contemporary lives, cultures, and ways of thinking. As a result, we must scrutinise some of the major symptoms of these produced spaces and systems that contribute to ecological degradation. While there are many examples of spatial injustice, such as gerrymandering voting districts, segregation, colonialism, security-obsessed surveillance, distribution, and private prisons, this book selects two of the largest spatially unjust environmental issues: the burning of fossil fuels and the consequences of climate change. One of the more effective ways to interrogate and challenge space related to environmental pasts and futures is through the arts and humanities.

The environmental humanities emerged in the early 2000s – along with the 'spatial turn' in the humanities (sometimes aligned with the geohumanities) – as a way to think about cultural and artistic approaches to ecological concerns through various disciplines, such as history, philosophy, film, geography, religious studies, music, media, and literature. Many academic and community-based programmes have adopted inclusive curriculums centred on the environmental humanities to harness transnational and transcultural voices in the global effort to marshal support for reducing the environmental crisis.[3] The environmental humanities combine a global community of scholars, writers, and artists working in science, culture, arts, and social sciences into a transdisciplinary method of study (see Chapter 2). Rather than reducing the humanities into sub-disciplines, as intended by the traditional organising structure of the Western university, scholars and artists in the environmental humanities gather collective tools and strengths for various disciplines and methodological frameworks to understand and combat ecological challenges.

Artistic production, as a major element of the humanities, assists us to think through difficult political and social blockages as symbolic acts underlining the environmental crisis, providing a frequently absent element from global conversations. The environmental philosopher and literary scholar Timothy Morton writes, 'Art can help us, because it's a place in our culture that deals with intensity, shame, abjection, and loss. It also deals with reality and unreality, being and seeming' (2010, 10). Morton emphasised elsewhere, while speaking on a panel discussion entitled 'Creativity in the Face of Climate Change' at the University of California, Berkeley in 2009, 'art as science and science as art' should intermingle to monitor and listen to the world.[4] An earlier event occurred at Oxford University in 2005. Hosted by the Environmental Change Institute, 30 scientists and 30 artists (some of which included Ian McEwan, Robert Macfarlane, Philip Pullman, Caryl Churchill, and Gretel Ehrlich) came together to brainstorm about how art and science might collaborate in the fight against climate change (Macfarlane 2005). Art, in this sense, serves as a type of scientific experiment of representation revealing answers to environmental and social problems.

The focus of the environmental humanities is not as anthropocentric as it may appear. Scholars, artists, and activists working in the field often underscore the distinctive power relations between human responsibility and hierarchy, decentring the human among other nonhuman (animal or posthuman) or biocentric urgencies. Likewise, the Anthropocene, defined as the industrialised 'age of the human' (1784–present) and discussed more in Chapter 3, acknowledges human capacity to create significant geophysical change of the Earth, but without elevating the human as the primary concern. While humans hold a certain planetary responsibility, they also fit into a matrix of planetary species and organisms affected by ecological change.

Drawing on the transdisciplinary commitment of the environmental humanities, I decided to examine various forms of literary and visual texts because confronting contemporary ecological degradation demands that we use many tools, including different technologies, formats, and aesthetics, to create and distribute ideas. For example, a filmpoem posted on an open-access website (Chapter 5) or a short story published by a traditional publisher (Chapter 9) both might underscore the same ecological issue, but in different ways and to distinctive audiences. How might a viral YouTube video confronting LEGO about their use and promotion of Shell oil in children's toys (Chapter 2) differ in approach than an avant-garde play about North Sea oil and gas in 1970s Scotland (Chapter 4)? Both address oil and culture as ecospatial forms of injustice, but both use different approaches to mobilise audiences. This broader approach has its obvious strengths and drawbacks, where focus must concede allowance for divergent and yet interlinking disciplines. In this way, the specificity of the subject and locations in this book drive the artistic responses to them.

To locate *Ecological Exile* in a particular space and place, I focus on writers, filmmakers, and media produced and set in parts of the United Kingdom (UK) and Ireland as case studies.[5] The reasons for locating this work in these specific geographies are twofold. First, from a pragmatic standpoint, a book of this magnitude must narrow its scope. Because these locations are in my area of expertise, they have logically become the concentration of my analysis. Second, both UK and Ireland – often referred to as the archipelago of the British Isles as a geographical referent extending into the North Atlantic and Arctic – offer a large range of widely consumed forms of anglophone literature, film, and media representations that feature key issues addressed in this book, namely the effects of oil and gas production and climate change.

Although the Global South deservedly remains a hotspot for critical geographical and ecological attention, the Global North is also relevant in planetary discussions about climate and energy, particularly in the northern peripheries discussed in this book, which are away from and function in opposition to the centres of commerce traditionally associated with the Global North, such as London, New York, or Paris. As postcolonial critics have demonstrated for decades, the expansion and success of the Global North resulted from centuries of exploitation and oppression of people, animals, and terrains in the Global South (Szeman and Boyer 2017, 2). Such an unjust scenario remains an enormous issue and concern. However, related oppression and systematic exploitation has occurred in the Global North by dominant power structures. Ireland, Scotland, and even rural parts of England, for example, continue to be occupied, colonised by polluting multinational corporations, but often sold out by their own central governments. This book largely addresses these locations in parts of Ireland, Scotland, Shetland, England, Wales, Greenland, and the North Sea that oppose economic and political centres attached to these otherwise outlying geographies. They are also locations in the Global North more affected by climate change (islands surrounded by coastlines) and subjected to the injustices of extractive industries (oil, natural gas, and refineries). Rather than serving as a reductive analysis of art as activism, my approach is to show how artistic production coming from and focused on these specific places of spatial injustice speaks to and for an issue, thereby becoming a 'socially symbolic act'. Instead of functioning only as social or historical, these places highlight a distinction between artistic production arising out of the places affected by spatial injustices and those literary or visual texts speaking to the same issue but from different and yet relatable places.

A guide to the book

This book is divided into three parts to provide coherence to an otherwise enormous set of topics. By focusing on injustices related to space, oil, and climate – all simultaneously material, abstract, and symbolic – I aim to highlight synergistic problems that are arguably the most significant environmental concerns facing the twenty-first century. Through organising this book around three interrelated and yet complex themes within the environmental humanities, I want to underscore how the modern fossil economy and related culture have contributed to our current climate crisis, creating ecospatial injustices in various places.

Individual chapters in the three thematic parts of the book provide examples of spatial injustices through literary and visual representations that challenge and promote how we can produce and construct the place-homes in which we live. Some of these creative examples serve as primary texts and receive significant analysis, which I outline in the chapter summaries below, while other literary or visual examples dispersed throughout the book serve as secondary texts that illustrate specific points. Many of the secondary texts are part of the 2015 series in *The Guardian* entitled 'Keep it in the Ground: a poem a day', and subtitled 'a climate change poem for today'. Curated by Carol Anne Duffy, the Poet Laureate of the UK, this series produced 20 poems as responses to climate change. Although I discuss this project more in Chapter 7, some of these poems are briefly examined throughout the book as examples of literary texts posted online as part of a cultural climate initiative. Some of the poets include Jackie Kay, Carol Anne Duffy, Theo Dorgan, Rachael Boast, and Gillian Clarke.

The opening chapters of each of the three parts of the book (1, 3, and 7) contain more extensive explanations surrounding the primary issues to provide context, while the remaining chapters of each of the three main parts offer closer analysis of literary and visual texts. In this sense, *Ecological Exile* does not strictly provide interpretations or close readings of literary and visual texts in each chapter. It also overviews for readers some of the practical and theoretical links between spatial justice and the environmental humanities, specifically through the effects of fossil fuels and climate change in culture.

Part I, 'Space', provides an introduction for two of the larger fields of study underlining the book – spatial in/justice and the environmental humanities – while it also analyses solastalgia and its useful application in space and place studies, particularly with regard to literary and visual culture. Chapter 1, 'Spatial in/justice and place', outlines a brief critical history of spatial and place justice in geography and social theory. Included throughout are methodologies by key theorists, such as Henri Lefebvre, Michel Foucault, David Harvey, Doreen Massey, Edward Soja, Iris Marion Young, Tim Cresswell, Yi Fu Tuan, and Edward Relph, who have contributed to constructing the paradigms of spatial justice and place studies. Offering relevant connections to this book's environmental focus, this first chapter primarily serves as a critical introduction for the book, but it also reinforces the overall argument and provides some literary and visual examples throughout.

Chapter 2, 'Solastalgia and the environmental humanities', begins by explaining the concept of solastalgia and how it serves as a specific way to understand spatial injustices related to place and environmental damage. This overview sets up the next section of the chapter, which explains the environmental humanities, along with an applied analysis of a Greenpeace UK viral YouTube video that confronts drilling in the Arctic entitled *LEGO: Everything Is NOT Awesome* (2014). The last section of the chapter investigates Eavan Boland's poem 'In Our Own Country' (2007), which pairs with Oliver Comerford's short film *Distance* (2004) as a unified filmpoem available on the open-access website 'The Poetry Project: Poetry and Art from Ireland'. Boland and Comerford through both visual and literary texts perceive social layers of place, with accumulated pasts, histories, and memories that all underscore place-home through the geocritical notion of stratigraphic vision.

Part II, 'Oil', demonstrates how oil, which is used as a shorthand name for fossil fuels more generally, is as much social and cultural as economic and infrastructural. Chapter 3, 'Petrospaces', discusses the petrocultures and the energy humanities, time-space compression of neoliberalism, and the Anthropocene within the period of modernity known as petromodernity. Approaching various oil-related injustices from a petrospatial perspective, this chapter surveys the links among culture, space, and injustice as a way to view how the production of fossil fuel spaces are represented in literary and visual texts, and in culture more widely. It ultimately considers how the time-space compression of neoliberalism within what can be called petromodernity has created spaces of placelessness and place-taking that contribute to spatial injustices and feelings of solastalgia.

Chapter 4, 'Speed of petrodrama', examines John McGrath's play *The Cheviot, the Stage, and the Black, Black Oil* (1973) about contested Scottish Highland histories during the North Sea oil boom. Through the flexible spatial form of agit-prop theatre, the play confronts ecologically damaging oil infrastructures from the multinational control of North Sea oil and gas by mobilising action through its theatrical form as a petrodrama. This chapter argues that *The Cheviot* exhibits the political speed of the fossil fuel economy staged within an improvisational theatre performance, which demonstrates how the acceleration of economic processes in spaces of global capitalism affect social lives and surrounding ecologies of place.

Chapter 5, 'Sullom Voe', explores how Roseanne Watt's filmpoem *Sullom* (2014) unsettles dominant histories in displacement in the Shetland Isles by reframing the narrative through poetry about place, thereby confronting environmental and spatial injustices within the history of North Sea oil in Scotland.

Watt frames the spaces of Sullom Voe in two primary ways: through the real and imagined body, and the multidimensional effect of the visual and poetic hologram. This bi-medial relationship (as an image-text) connects the poem (as a literary text) to the film (as a media text). *Sullom*'s musical score is an additional element that creates an anti-aesthetic effect, ironising petrochemical advertisement campaigns produced by energy companies such as Suncor Energy's *See What Yes Can Do* (2013).

Chapter 6, 'Pipelines of injustice', investigates the Irish documentary film *The Pipe* (2010) by Richard O'Donnell. This chapter surveys the injustices of pipelines and their social and spatial effects on small communities (micro-minorities) and the places they are often built (micro-geographies). *The Pipe* serves as a visual and cultural narrative of opposition, capturing the social process of storytelling as resistance by the people of Rossport, Ireland. The beginning of the chapter partly draws on comparative analyses from Donal O'Kelly's play about Nigerian and Irish oil and gas, *Little Thing, Big Thing* (2014), which references Ken Saro-Wiwa's well-documented literary activism of the 1990s in response to Shell's domination of the Niger Delta. The documentary addresses what eventually leads to solastalgia – the people of Rossport are exiles relegated to remain in their home, but their mobility and social support in the community have been compromised through spatial and social displacement.

Part III, 'Climate', contends that anthropogenic (human-caused) climate change remains a spatial justice issue as much as an environmental one. The people who suffer the most from consequences of flooding, lack of drinking water, and diminishing food supplies are not the people who cause or maintain the climate crisis. Chapter 7, 'Climate injustice', overviews the climate science and the conceptual challenges of addressing climate inaction. The final section of the chapter examines an Anglo-Danish documentary released in the UK and Denmark entitled *Village at the End of the World* (2012). It was made by the filmmaker Sarah Gavron, who recently directed the acclaimed social justice narrative film *Suffragette* (2015), and co-director/cinematographer David Katznelson. *Village* illustrates unjust circumstances of climate change, but it does so as a story of climate injustice through the voices and experiences of the Inuit of Greenland. The film confronts contemporary problems that largely revolve around changing economics, climate change, and decolonisation around the planet.

Chapter 8, 'Cli-fi', explores the literary genre or style of writing called 'cli-fi' – otherwise known as climate (change) fiction. Cli-fi not only draws out explicit concerns with environmental futures (catastrophe or apocalypse), but also challenges the systems that have produced the climate crisis. This chapter specifically investigates the spatial elements of Ian McEwan's cli-fi novel *Solar* (2010) through Georg Lukács' notion of transcendental homelessness, an approach that pushes the interpretation of the novel beyond the protagonist's notable and at times exaggerated foibles and into the realm of his fragmented self in the modern environmental crisis.

Chapter 9, 'Irony of catastrophe', explores how the use of irony as a stylistic tool for persuasion about climate change in China Miéville's 'Polynia' (2015) and Kevin Barry's 'Fjord of Killary' (2012) underscores spatial elements that resonate with themes of injustice resulting in solastalgia. Both conceptually and stylistically, the short story form allows writers to interrogate specific themes through direct and poignant motifs as a formal totality. Within this literary form, one predominant theme is clear: short stories, specifically 'Polynia' and 'Fjord of Killary', consistently establish an awareness of place. This chapter also serves as an apt conclusion for the book by addressing 'end of the world' catastrophe narratives through comic forms of short fiction and what that means moving forward in a potential age of increased calamity.

If space is socially constructed, often through accounts of hierarchical power, then how do forms of literary and visual texts change that narrative, and in turn our perception and ontological understanding of space? How do texts and images construct new epistemologies and ontologies about space? How do they ease or inform of the tension between environments shaping us versus our attempt to shape our environment to suit our needs? This book ultimately explores how various writers, poets, documentary filmmakers, and visual artists explore and challenge injustices in built and non-built spaces to support a socially and environmentally sustainable future.

Notes

1 There are multiple uses of the term 'disciplinarity'. A brief definition would support a constant interaction and overlap of various methodologies. This, however, falls short of the nuances assigned to various terms. Interdisciplinary approaches attempt to integrate methods from various disciplines using real synthesis of each approach; transdisciplinarity creates a unity of methodological frameworks that range beyond individual ones; multidisciplinary approaches are from people in different disciplines who work together, drawing on each person's expertise as a collective; cross-disciplinarity looks at one discipline from perspectives of another; intra-disciplinary methods are rooted in one single discipline. While this book is largely interdisciplinary and transdisciplinary, it also blends some of these other concepts at various points.

2 Elsewhere, Jameson introduces the concept of 'cognitive mapping' as a way to describe 'a pedagogical political culture' seeking 'to endow the individual subject with some new heightened sense of his place in the global system' (1984, 92). Capitalism, referring to the 'global system', is a large part but not the only force producing space and responses of artistic production within it. In addition, cognitive mapping underlies the subject's relationship to simultaneously immediate urban and global spaces. Cognitive mapping serves as a useful tool for political interpretations of aesthetic practices developed within capitalist culture, and particularly those in urban spaces, some of which overlap with spatial injustices of oil and climate change, and some of which do not (see Chapter 3).

3 Some useful overview volumes of the environmental humanities include: Frawley and McCalman (2014), DeLoughrey et al. (2015), Adamson and Davis (2016), and Iovino and Oppermann (2016).

4 See 'Creativity in the Face of Climate Change: The Role of the Humanities in Awakening Societal Change'. Berkeley Institute of the Environment, 2 February 2009. Accessed 2 July 2016. www.youtube.com/watch?v=in5F3OfbtUA.

5 Two exceptions exist in this book, both of which are produced in the UK but focus on the Arctic: *LEGO: Everything Is NOT Awesome* (Chapter 2) and *Village at the End of the World* (Chapter 7).

Bibliography

Adamson, Joni, and Michael Davis, eds. 2016. *Humanities for the Environment: Integrating Knowledge, Forging New Constellations of Practice*. London: Routledge.

DeLoughrey, Elizabeth, Jill Didur, and Anthony Carrigan, eds. 2015. *Global Ecologies and the Environmental Humanities: Postcolonial Approaches*. New York: Routledge.

Frawley, Jodi, and Iain McCalman, eds. 2014. *Rethinking Invasion Ecologies from the Environmental Humanities*. London: Routledge.

hooks, bell. 1990. *Yearning: Race, Gender, and Cultural Politics*. Boston, MA: South End Press.

Iovino, Serenella, and Serpil Oppermann, eds. 2016. *Environmental Humanities: Voices from the Anthropocene*. London: Rowman & Littlefield International.

Jameson, Fredric. 1981. *The Political Unconscious: Narrative as a Socially Symbolic Act*. Ithaca, NY: Cornell University Press.

——. 1984. 'Postmodernism, or the Cultural Logic of Late Capitalism'. *New Left Review* 146: 53–92.

Kearns, Gerry. 2014. 'Introduction'. In *Spatial Justice and the Irish Crisis*, edited by Gerry Kearns, David Meredith, and John Morrissey, 1–16. Dublin: Royal Irish Academy.

Lefebvre, Henri. 1991 [1974]. *The Production of Space*. Translated by Donald Nicholson-Smith. Oxford: Blackwell.

Macfarlane, Robert. 2005. 'The Burning Question'. *The Guardian*, 24 September. Accessed 3 May 2016. www.theguardian.com/books/2005/sep/24/featuresreviews.guardianreview29.

Mayer, Sylvia. 2006. 'Literary Studies, Ecofeminism, and the Relevance of Environmentalist Knowledge Production in the Humanities'. In *Nature in Literary and Cultural Studies: Transatlantic Conversations on Ecocriticism*, edited by Catrin Gersdorf and Sylvia Mayer, 111–28. Amsterdam: Editions Rodopi BV.

Morton, Timothy. 2010. *The Ecological Thought*. Cambridge, MA: Harvard University Press.

Parmar, Pratibha. 1990. 'Black Feminism: The Politics of Articulation'. In *Identity: Community, Culture, Difference*, edited by Jonathan Rutherford, 101–26. London: Lawrence & Wishart.

Smyth, Gerry. 2001. *Space and the Irish Cultural Imagination*. Basingstoke, UK: Palgrave.

Soja, Edward. 2010. *Seeking Spatial Justice*. Minneapolis, MN: University of Minnesota Press.

Szeman, Imre, and Dominic Boyer. 2017. 'Introduction: On the Energy Humanities'. In *Energy Humanities: An Anthology*, edited by Imre Szeman and Dominic Boyer, 1–13. Baltimore, MD: Johns Hopkins University Press.

Trexler, Adam, and Adeline Johns-Putra. 2011. 'Climate Change in Literature and Literary Criticism'. *WIREs: Climate Change* 2: 185–200.

Part I

Space

1 Spatial in/justice and place

In an attempt to provide a background of spatial justice that complements approaches in the environmental humanities for the following chapters of the book, this opening chapter of Part I, 'Space', outlines a brief critical history of spatial and place justice in geography and social theory. Chapter 2 further demonstrates how a spatial theories of justice might fruitfully apply to scholarship and writing in the environmental humanities. Included throughout, I highlight some key figures, such as Henri Lefebvre, Michel Foucault, David Harvey, Doreen Massey, Edward Soja, Iris Marion Young, Tim Cresswell, Yi Fu Tuan, and Edward Relph, who all have contributed to constructing the paradigms of spatial justice and place studies. Their writings do more than outline spatial theory; they provide a foundation for this book's environmental focus. This first chapter, the first half of Part I, primarily serves as a critical introduction for *Ecological Exile*. It also reinforces the overall argument and provides some literary and visual examples throughout. The main objective here is to establish how spatial and environmental justice anticipate solastalgia as it relates to place (see Chapter 2), providing the groundwork for the remaining chapters of the book.

Spatial theories of justice

One universal principle underlying spatial theories of justice is that wherever there are geographies, there are social injustices. By this logical perspective, spatial justice is a collective and universal issue whether it is acknowledged or not. Edward Soja, the urban planner, postmodern geographer, and promoter of spatial justice as a way of addressing many social issues, explains, 'Space is not an empty void. It is always filled with politics, ideology, and other forces shaping our lives and challenging us to engage in struggles over geography' (2010, 19). Concepts of space and justice responding to unjust geographies have existed in various documentable forms dating back to at least the social philosophies of ancient Athens. Spatial justice has, however, only been significantly theorised and developed over the last 50 years in the wake of civil, financial, and environmental distresses. In these contemporary contexts, spatial justice illuminates how access to social goods and services depend upon where one lives or works (Smith 1994). People's ethnicity, class, or gender, in combination with where they find themselves in space, may affect how they are treated or their right to participate equally in society. Another purpose of spatial justice is to identify who manages public and private space and in what ways those controllers promote oppression or equality. To this end, there are numerous examples of how spatial justice functions in many global contexts.

Rather than functioning as an amorphous fixed background (i.e. abstract or inert), the notion of space in critical spatial theory covers a range of interconnected issues for humans and nonhumans alike, which, for the purposes of this book, often overlap with environmental justice (Soja 2010, 4). Space not only constitutes physical or material form; it is also socially produced and constructed as an idea or concept, affecting urban economies, environmental policies, and human rights, as well as forms of creative output in the arts and humanities. For example, mismanagement of urban space – such as placing industrial waste sites near economically depressed neighbourhoods, gentrifying urban zones, or redrawing voting districts for political gain – actively generates instances of inequality and exploitation through a variety of forms of oppression and discrimination.

Referring to both just and unjust circumstances of space (hence the reversible and reflexive term 'in/justice'), spatial justice was developed and theorised as a specific approach in Soja's book *Seeking Spatial Justice* (2010). In it, as explained more throughout this section, Soja aimed to expand the parameters of social justice as spatial phenomena by emphasising that space and geography remain an important element to any discipline engaging with social and cultural formations (2010, 4). While Soja's book serves as the most extensive study on the subject to date, he is only one of many theorists before him who have explored the underling motivation for developing the term spatial justice – that is, how social struggles over territories and geographies link to injustices.

Since the 1960s, geographers and social philosophers have confronted territorial struggles by arguing that spatial dynamics, not only historical or social ones, should be central in debates about improving society.[1] Critical perspectives that challenge oppressive and undemocratic spatial practices largely date back to the work of French urbanist Henri Lefebvre and philosopher Michel Foucault, among others, when in the 1960s and 1970s they wrote and lectured about the fundamental relationship between space and power in society (Lefebvre 1968, 1991; Foucault 1980, 1982, 1986). The Marxist geographer David Harvey built on this work in the 1970s and fundamentally cemented spatial studies into critical discourse by applying social and moral philosophy to issues related to geography and environments.

In *Social Justice and the City* (2009 [1973]), his book on the relationship between urban space and social processes that still resonates today, Harvey deduces that principles of social justice had significant relevance for the application of spatial and geographical approaches to urbanisation (2009, 9). He argues the 'distinction between social processes and spatial form is always regarded as artificial rather than real'. But, he then suggests, 'Spatial forms are there seen not as inanimate objects within which the social process unfolds, but as things which "contain" social processes in the same manner that social processes *are* spatial' (Harvey 2009, 11; original emphasis). Harvey, at the time, called this dynamic the 'social-process-spatial-form theme', insisting upon the need to understand the ways in which human activity in society creates a need for distinct spatial concepts and understanding (2009, 14).

Harvey drew from and expanded upon the idea of 'territorial justice' originally coined by the Welsh social planner Bleddyn Davis in his book *Social Needs and Resources in Local Services* (1968), which reasoned the allocation of public services to territories must meet specific social needs. For Harvey, territorial justice extends beyond equal distribution of resources, although he acknowledges that it forms the basis of distributive justice. Harvey concluded that understanding processes of justice are not enough (in reflection of Rawls' *Theory of Justice*); instead, he argued that territorial injustices, what he calls 'territorial social justices', emerge from unjustly produced power in social spaces, specifically through urban development and income distribution (2009, 99–101). Territorial justice serves as a fundamental principle of spatial justice, but perhaps it falls short in the ways it offers one component of a larger theory categorising various conditions of injustices that root from spatial perspectives. Territorial social justice relates exclusively to forms of distribution in relation to social constructions of power, whereas spatial justice provides a wider range of applications to issues such as unjust racial, gender, or environmental conditions caused by space, which all might include but are not limited to territorial justice.

Other geographers such as Doreen Massey (1994) and Derek Gregory (1994), in addition to Soja (1989) and Harvey (1990, 1996a, 2009), later pointed out how social processes are also spatial in forms of environmental, postcolonial, gender, or class injustices. This productive dialogue between geography and social theory is commonly known as the 'spatial turn', or a way of reimagining social and historical perspectives through a spatial lens. These spatial perspectives were additionally rooted in the 'cultural turn' that took place within the discipline of geography beginning in the 1970s and expanding in the 1980s and 1990s. Both 'turns' (spatial and cultural) are broadly drawn from British social geographers interested in issues of space and power relations and American geographers concerned with symbolic productions of space (Scott 2004, 24). A primary objective of critical geography is to examine the relationships between humans and environments, and how these interactions shape the debates across various disciplinary boundaries (Duncan et al. 2004, 3). Geography, as Derek Gregory outlines in *Geographical Imaginations* (1994), 'is not confined to any one discipline, or ever to the specialised vocabularies of the academy; it travels instead through social practices at large and is implicated in myriad topographies of power and knowledge' (11). The spatial turn primarily resulted from cultural and human approaches to geography, modes that have migrated to many other forms of cultural and literary criticism in the humanities, and which drew from aspects of cultural theory, such as poststructuralism, feminism, postmodernism, and postcolonial theory (Scott 2004, 24).

The early fusion of cultural and spatial theories in the context of justice appears in the work of Foucault. He presented another way of critiquing issues of knowledge and power in culture and society, but he did so through a uniquely historically and philosophically informed viewpoint of space (Soja 1989, 31). Soja maintains:

> contributions of Foucault to the development of critical human geography must be drawn out archeologically, for he buried his precursory spatial turn in brilliant whirls of historical insight. He would no doubt have resisted being called a postmodern geographer, but he was one. (1989, 16; see also 1996, 144–63)

Archaeology is a suggestive term here because it implies a balanced proportion of the historical, social, and spatial measured out through stratigraphic layers of existence. In this way, time and space are mutually interdependent, even though notions of space have received much less critical attention and practical application.

Foucault famously stated in a lecture entitled 'Of Other Spaces / *Des espaces autres*', which was given to a group of architects on 14 March 1967 and posthumously published, that discourses of the nineteenth century were dominated by time, history, and evolution, whereas thinkers in the twentieth century finally began to examine ideas about space as they related to the modern and postmodern condition (1986, 22). He later admitted, 'Geography must indeed necessarily lie at the heart of my concerns' (Foucault 2007, 182). Foucault's statements about space challenged the disciplinary tendency to privilege historical time; he instead invited us to think more critically about how space substantially informs both historical circumstances and social practices.

For these reasons, contemporary approaches to spatial justice take as a starting point the late 1960s, when perspectives combining the interplay among space, geography, and social justice were being developed and theorised in the wake of political unrest in the student and labour strikes of Paris in May 1968. These mass protests both acknowledged and confronted spatial injustices, particularly resisting urban renewal projects that would displace working class populations out to the margins of Paris in fear they might maintain an electoral majority (Soja 2010, 100). Paris is not my focus, however. It is more of a critical point of departure or the spatial zeitgeist for what is ahead in the overall historical logic of this book, which recognises the shift whereby formulations of spatial justice began to take root.

Critical responses to unjust spaces that began in the 1960s were largely due to a century and a half of privileging social historicism as the *only* way to examine territorial struggles. The aim was to challenge the idea that space is no longer considered a flat or one-dimensional way of documenting geography, as in traditional forms of cartography often influenced by oppressive empires and feudal societies, or as an absolute *thing* in itself. The issue of space has become ontological and epistemological – intimately connected to the ways we know and exist in the world, often addressing questions about the experience of being in and understanding space.

The more difficult question to answer might be what exactly is space? The problem with this question is that space is relational as well as material, what Harvey theorised as a relationship between and within objects (2009, 12). And so, we must then ask how do different human and cultural practices use and conceptualise space? In a 2014 interview with the geographer John Morrissey, over 40 years after *Social Justice and the City* was published, Harvey offers another important and yet similar question for consideration: 'what is the social process which is actually producing these geographical patterns in which you see the embeddedness of injustice in the landscape?' (Morrissey 2014, 213). This question implies a social-cultural nexus that might also be explored through the ways creative/artistic production respond to spatial injustices.

A social approach to spatial justice is profitably expanded in the cultural realm of the arts and humanities, where creative responses in literature and filmmaking, to cite only two examples, might clarify how real and imagined forms of injustices related to space are represented in narrative, setting, dialogue, or embedded in the social context of a text. Indeed, spatial theorists often cite literature, film, or paintings to illustrate otherwise abstract concepts. Print and visual texts aim to communicate through various constructions of spatial forms. In particular, the written word contains properties 'from the flux of experience and fixes them in spatial form'. As Harvey goes on to explain, any system of representation, of which I would argue defines the essence of creative practice in the arts and humanities, functions as a spatial process because it isolates the flow of experience, thereby distorting what it is attempting to represent (1990, 206).

Digging a bit deeper, critical spatial thinking draws upon three main principles. First is the concept of 'ontologies of being', which underscores that we are all spatial beings as much as we are social and historical in the ways we know and exist in the world. Second is the 'social production of space', or the idea that space can be socially changed because it is socially constructed and produced based upon a concept or idea, affecting economies, politics, and cultural/creative production. Third is what Soja has labelled the 'socio-spatial dialectic', which demonstrates how the social and spatial are mutually dependent and shape each other in various ways (1989; 2009, 2). These concepts were all originally theorised by Lefebvre in *La Production de l'espace* (1974), later translated as *The Production of Space* (1991), and, along with Harvey's notion of territorial justice, they are building blocks for spatial theories of justice later developed by Soja in *Seeking Spatial Justice* (2010).

Spatial justice arises out of a new spatial consciousness, what Harvey once called the 'social-process-spatial-form' and what Soja has termed the 'socio-spatial dialectic'. It is the 'idea that there exists a mutually influential and formative relation between the social and the spatial dimensions of human life, each shaping the other in similar ways' (Soja 2010, 4). The concept of spatial justice underscores how geography as a study and lived experience is a socialised process largely associated with cultural formations. Soja's socio-spatial dialectic theory contends that we are spatial beings and engage with space in a reciprocal, two-way process: from the point of birth, we attempt to shape our environment to suit our needs; at the same time, the environment is constantly shaping us (1989, 76–78; 2009, 2). Soja echoes Lefebvre's claim mentioned below that while humans are producing space, they are in turn constantly produced by it. This mutually dependent process of socio-spatial relationality explains the correlation between our needs and how geography shapes us. The relationship between processes of production and products of social spaces in our daily lives is specifically addressed by Lefebvre.

While Lefebvre outlines approaches to daily experience in urban spaces throughout many of his writings, his book *The Production of Space* (1991) acutely explores produced knowledge about space.[2] In it, Lefebvre circuitously incorporates philosophical theories from Karl Marx, Friedrich Nietzsche, Georg Hegel, and Sigmund Freud, to Julia Kristeva, Jacques Derrida, and Roland Barthes, as well as creative and aesthetic encounters with artistic forms of literature and music. Lefebvre believed that life should be lived as a project, an unfolding process of both theory and practice. Creative forms of literary and visual culture aptly represent the lived and material elements of the lived project. Harvey writes in the 'Afterword' to the 1991 translation of *The Production of Space* that Lefebvre's 'belief in the animating power of spectacle, of poetry, and of artistic practices became crucial in informing Lefebvre's attitude towards and active participation in the revolutionary movements of the 1960s' (1991, 426). To this effect, Lefebvre understood how artistic and creative practices reproduced quotidian daily life, which in turn had the benefit of provoking social action by constructing new mental and lived spaces in society. Lefebvre maintained that the 'form of social space is encounter, assembly, simultaneity' (1991, 101). When speaking about space, Foucault similarly affirmed, 'we are in the epoch of simultaneity: we are in the epoch of juxtaposition, the epoch of the near and far, of the side-by-side, of the dispersed' (1986, 22). *The Production of Space* operates as a key text to later applications of spatial justice because of its understanding of simultaneous lived spaces within social structures.

Thinking about space as more than just a physical form or a neutral abstract background invokes Lefebvre's well-documented idea that space is socially produced: 'that (social) space is a (social) product' (1991, 30). Rather than serving as a tautology, reinforcing the seemingly circular equation that social space is a social product, it instead underscores the three-way relationship among social, mental, and physical space. The aggregate of these three elements form ontologies, or ways of being in the world, which is to say what humans produce also shapes and produces them. The production process of social systems (e.g. economy, state, or education) is inseparable to and the same as the product (arts, media, culture). He further explains, 'Though a *product* to be used, to be consumed, it [space] also is a *means of production*; networks of exchange and flows of raw materials and energy fashion space and are determined by it' (1991, 85; original emphasis). We can to a large extent make, shape, and construct the real and imagined geographies in our lives. Spaces are not only inert or unchangeably 'real' or material monoliths in our lives; rather, they are also perceived or 'imagined' socially. Injustices occur when those in positions of power produce space in limited and limiting ways, serving the interests of the privileged few instead of the collective. These people are, of course, typically those who do not have to live according to those limiting and limited social constructions.

Lefebvre's writings on space pose two widely interconnected questions: what are the spaces that those in power want to construct and to what ends? What forms of alternative space might be invented to serve as acts of resistance and opposition? (Conley 2012, 12). The first question is historical while the second is more imagined. Both, however, assume space is multidimensional; it is subtle and intricate. It is the philosopher's task, he suggests, to engage the collective consciousness about opportunities and transformations to combat alienation and promote generative social experiences. These experiences promote critical, creative, and cultural awareness that underline Lefebvre's '*metaphilosophy*', or a way to critique the emergence of other unjust spaces.

As a guide to understanding how space is socially produced, Lefebvre outlines a triad of social space: spatial practice (the experienced/perceived), representations of space (the conceived), and representational spaces (the imagined). Spatial practice, or 'perceived' space, suggests 'a society secretes that society's space; it propounds and presupposes it, in a dialectical interaction; it produces it slowly and surely as it masters and appropriates it' (Lefebvre 1991, 38). It is a way of examining physical space during what Lefebvre calls 'neocapitalism' (i.e. neoliberalism or late capitalism), particularly in traditional geography when material spaces are perceived, analysed, and appropriated on a macro level to obtain power. Spatial practice deciphers the routines, routes, and networks that link to both private and public life within perceived spaces.

The second approach called representations of space builds upon the idea of spatial practice. Rather than viewing space as perceived or what is one-dimensionally represented in spatial practice, he inverts it as 'conceived' space, also referred to as 'imagined'. For Lefebvre, representations of space are 'conceptualized space, the space of scientists, planners, urbanists, technocratic subdividers and social engineers . . . all of whom identify what is lived and what is perceived with what is conceived' (1991, 38). Whereas spatial practice focuses on real spaces, representations of space emphasise imagined spaces, which include codes of knowledge and signs as ways of understanding experienced places often articulated by experts.

The third and last illustration of Lefebvre's triad of social space labelled representational spaces, or 'lived' space, is:

> directly *lived* through its associated images and symbols, and hence the space of 'inhabitants' and 'users', but also of some artists and perhaps of those, such as a few writers and philosophers, who describe and aspire to do no more than describe. (1991, 39; original emphasis).

Lefebvre goes on to explain that representational spaces 'tend towards more or less coherent systems of non-verbal symbols and signs' (1991, 39). How do people ultimately 'live' in real (perceived) or imagined (conceived) spaces? They must imagine and construct new meaning and possibility, and remake spaces to support a sustainable society.

Lefebvre's 'perceived-conceived-lived triad' of space explicated in *The Production of Space* is a useful guide for many implicit and explicit reasons in the following chapters (1991, 40). The main purpose in discussing it here at some length is to demonstrate the 'real and imagined' relationship of social space (including the 'lived' in a simultaneous triad). These underlining frameworks of spatial injustice that can be applied to systems of ecology as part of larger social spaces can be used by writers and filmmakers, as much as geographers or philosophers. Artistic forms are both socially produced and products that incorporate Lefebvre's spatial triad. Tuscan landscape painters, as he cites, developed conceived space that also incorporated the perceived real spaces of landscapes and examples of lived experience within these spaces (Lefebvre 1991, 41). Novels or films are no different. They produce the real and imagined spaces that reflect social conditions while also creating new ones through counter-narratives reconstructed as lived experience.

Conflicts arise not just out of diverse subjectivities of experience in social spaces, but largely because of varying objective material realities of social life experienced in time and space (Harvey 1991, 205). As Foucault and later the postcolonial theorist Edward Said have pointed out, space can also be oppressive – through unjust distribution of resources and land or lack of health services based upon ethnic status or socio-economic factors. Said believed that none of us are separate from the spaces in which we live, and thus none of us are 'completely free from the struggle over geography'. Said's statement resembles Soja's socio-spatial dialectic, where space shapes us as we constantly attempt to shape the spaces around us. Thinking of spaces as imagined and socially constructed, and therefore highly influential in everyone's lives, Said goes on to explain in *Culture and Imperialism* that the 'struggle is complex and interesting because it is not only about soldiers and cannons but also about ideas, about forms, about images and imaginings' (1993, 7). Reproducing spaces comes through altering the imagination and values as much as material forms. The real and imagined elements of spatial production are on display here in Said's quote, with the imagined elements of the arts and humanities as crucial factors to understanding and reproducing space.

In addition to Foucault and Said, the feminist geographer Doreen Massey has argued throughout her prolific career that because westernised spaces are built upon masculine principles of patriarchy, it remains difficult for women and minorities to obtain and maintain power, the result of which would offer transformative social reproductions of space. Massey maintains that 'geography matters to gender' because constructions of each in society influence cultural formations that are 'deep and multifarious' (1994, 177). Speaking about the politics of race and gender, the feminist theorist bell hooks explores the injustices of 'space and location' that 'evoked pain' for the marginalised. For hooks, in order for a personal and political 'artistic evolution' to occur, one must 'face ways' space and location are 'intimately connected to intense personal emotional upheaval regarding place, identity, desire' (1990, 146). Marginality is a space occupied by the oppressed, exploited, colonized people; but 'telling of a sense of place' through 'artistic and literary practice' can interrupt and transform such margins (hooks 1990, 146, 152). As hooks explains, '[s]peaking from the margins' is an action of spatial 'resistance' (1990, 152). Both gender and race exemplify how spaces create injustices, which reinforce similar outcomes in environmental justice. In this way, spatial justice draws on fundamental principles in the humanities by utilising images and the imagination to both control spaces and resist injustices related to space and society.

Social and spatial theory distinguish between concepts of *structure* and *agency*, emphasising the relationship between social systems of power and everyday lives of individuals and communities. Whereas Harvey, Foucault, and Lefebvre investigated systems of power as a spatial phenomenon, Soja, Said, Massey, and hooks extended their analysis into realms of agency to understand how attitudes, beliefs, and emotions might also reflexively respond to social systems of power (i.e. the socio-spatial dialectic) (Dear 2011, 6). Human existence on the planet remains highly dependent upon ontologically formed social, historical, and spatial circumstances, but spatial dynamics are often overlooked in this triad. However people might view their world, spatial thinking, in addition to social and historical methods, can enhance it (Soja 2013). There are more equitable or just ways of producing space through a distribution of resources, providing equal opportunities, or providing a plurality of voices. These are all some reasons to think about why we might turn to a more spatial perspective in society and, more specifically for the purposes of this book, reflect upon it when reading works of literature, watching films, or ingesting media culture.

Although the concept of justice differs depending upon social contexts and in different communities, it can nevertheless be understood somewhat universally in democracies where justice exists as a sociocultural construct. Largely defining 'justice' as 'the first virtue of social institutions', John Rawls' foundational text *A Theory of Justice* (1971) emphasises equitable forms of distributive justice in liberal democracies. For Rawls, justice holds two fundamental principles: first, everyone accepts that others undertake principles of justice; second, social institutions are designed to satisfy these principles (1971, 3). Rawls has been critiqued both from the right and left. Conservative critiques focus on Rawls' largely socialist position. Progressives argue that Rawls' theories of distributive justice do not go far enough and exist in a vacuum, devoid of historical, social, or spatial context (Soja 2010, 77; Young 2011, 23).

In *Justice and the Politics of Difference* (2011 [1990]), the political philosopher Iris Marion Young challenged and then added to Rawls' definition by claiming that theories of justice need to take into account specific geographical, institutional, or social influences. According to Young, 'social justice', as an idea and practice, 'requires not the melting away of differences, but institutions that promote reproduction and respect for group differences without oppression' (2011, 47). Injustice, in other words, equates to oppression. Unequal distribution or development may be a symptom of oppression, but it is not the cause of injustice. Young roots justice within social movements (and groups) to provide concrete examples of domination and oppression rather than relying on distributive injustice and its sole focus on wealth and income. In doing so, she classifies the 'five faces' of oppression/injustice: exploitation, marginalisation, powerlessness, cultural imperialism, and violence (Young 2011, 39). These five faces of oppression underline the environmental injustices discussed in the following chapters of *Ecological Exile*. While Young does not adopt a spatial perspective, she explains oppression in ways that overlap with it. Rather than only focusing on outcomes, I want to also draw on the social and spatial approaches to highlight causes or productions of environmental injustice that result in forms of oppression constituting solastalgia (see Chapter 2).

Drawing on the notion of social justice – producing substantive equality out of spatial and environmental oppression – this book does not task itself with the immense difficulty of defining justice as a comprehensive subject and theory or tracing its many legal debates. Rather, it documents how the following chapters address various understandings of justice based upon particular cultural and spatial perceptions and contexts. An internal logic exists when thinking through justice. Everyone deserves basic human rights. Harvey simplifies justice as 'the universal right for everyone to be treated with dignity and respect' (1996b, 94). The issue is that spaces have been reproduced to favour specific populations over others, and this outcome reinforces and normalises injustice. Fair and equitable laws governing societies do stand in stark contrast to unjust examples, but altering socially constructed systems that benefit the privileged few while oppressing the majority remains difficult to achieve without ways of perceiving different futures and altering values to support this shift.

Cultural and social differences exacerbate definitions of justice. The geographies covered in *Ecological Exile* are governed by westernised parliamentary democracies in the UK and Ireland. This distinction does not preclude injustices from occurring. Rather, the geographical and social peripheries in which the print and visual texts considered in this book capture environmental injustice and oppression under the British Parliament or the Irish Dáil (Assembly of Ireland). Justice, as a link to the spatial and environmental, is largely defined as seeking substantive equality and protection from planetary destruction on local, regional, and global scales. Many global justice systems, even in the democracies of Britain and Ireland, fail to account for differences between universal and localised rights, especially in terms of unfair spatial productions that favour specific groups of people by redistributing space, wealth, and essential services unevenly or by creating widespread social or geographical oppression. These outcomes generate spatial and environmental injustices. This is why theories of spatial justice, as well as place justice (discussed in the final section of this chapter), have been employed by geographers and social theorists. Artistic production in these areas also confront spatial injustices, albeit not so explicitly, hence the reason for this book.

Critical works

Prior to Soja's *Seeking Spatial Justice*, there were only a few published instances of the term 'spatial justice', all of which were employed by geographers (O'Laughlin 1973; Pirie 1983; Flusty 1994; Bromberg et al. 2007). Around the time *Seeking Spatial Justice* came out in 2010, a new online French journal appeared in 2009 entitled *Spatial Justice / Justice spatiale*. Up to this point,

Soja was the only scholar to base an entire book-length study around the concept of spatial justice (emphasising the importance of that term), even though some of the aforementioned geographers and philosophers in this chapter had already alluded to forms of spatial justice in other ways (i.e. territorial justice, urbanisation of justice, and struggle over geography). The following literature review outlines specific offerings crossing the spatial justice and environmental humanities divide, some more explicitly than others, beginning with Soja's pioneering book.

Soja's objective in *Seeking Spatial Justice* is both direct and broad: 'to stimulate new ways of thinking about and acting to change the unjust geographies in which we live' (2010, 5). He justifies the importance of critical spatial thinking and then builds his theory of spatial justice by working through three practical categories: exogenous geographies and the political organisation of space, endogenous geographies and spatial discrimination, and mesogeographies of uneven development. Exogenous geographies consider 'top-down' examples of territorial and hierarchical spatial injustices that include gerrymandering, apartheid, security-obsessed surveillance, and public-private spaces. Endogenous, or 'bottom-up', geographies involve distribution and location, such as unequal social distribution, racial discrimination (segregation), and environmental degradation. Mesogeographies underscore the unjust relationships between local and global uneven development of industrialised and developing countries, or what is currently referred to as the Global North and South. *Seeking Spatial Justice* concludes with a case study of spatial justice in action: the Los Angeles labour movement. Soja's book remains the key text of spatial justice and the reason for it receiving so much attention in this chapter and throughout *Ecological Exile*. It serves as a critical guide and pragmatic overview for future projects, and Soja intentionally leaves openings for further research to build upon this work. Exponential instances of global spatial injustices exist, even if they have not received much attention in disciplines outside of geography. My aim is to take up one of these gaps about how artistic production as part of the humanities might confront or affirm forms of spatial injustices.

Since the publication of *Seeking Spatial Justice*, two other books explicitly focused on spatial justice that cross the sociocultural divide somewhat into the humanities have appeared. Gerry Kearns, David Meredith, and John Morrissey edited the volume *Spatial Justice and the Irish Crisis* (2014) as a response to Ireland's financial and social crisis after the 2008 global economic crash. Not as theoretical as Soja's *Seeking Spatial Justice*, and arranged as a collection of essays from a variety of scholars in geography and the social sciences, *Spatial Justice* offers a practical examination of specific geographical inequalities and injustices that have occurred in contemporary Ireland leading up to and following the crash. The book targets geographers and students of geography focused on spatial concerns in Ireland related to finance, inequality, planning, and identity. The volume is worth mentioning here because it similarly addresses spatial, environmental, and place justice. While *Spatial Justice* remains the main geographical book-length contribution to this debate in Ireland, which informs some of my analysis in the environmental humanities, it focuses on policy and infrastructure more than cultural and artistic approaches to space, justice, and environments.

Andreas Philippopoulos-Mihalopoulos' *Spatial Justice: Body, Lawscape, Atmosphere* (2015) serves as the most recent and theoretical book-length study to examine space and justice, but it is situated in a legal context. It contends that all forms of justice are spatial in some way, whether constituted through written or oral law, as well as embodied social and political manifestations. As a scholar of law, Philippopoulos-Mihalopoulos navigates the legal terrains associated with 'law's spatial turn' in what he has termed 'spatiolegal thinking' (2015, ix). He examines the spatial relationships between bodies (as a posthuman concept including human, nonhuman, natural, material, elemental, or technological), and the ways they relate to each other in society. For Philippopoulos-Mihalopoulos, spatial justice is the consequence of struggling 'bodies' trying to occupy the same space at the same time, and this is demonstrated through two specific paradigms: *lawscape* and *atmosphere* (2015, 3; original emphasis). *Spatial Justice* offers an insightful theory-based approach to legal practice through social issues that relate to myriad forms of posthumanist bodies. The book's focus would suggest that it might appeal more to readers in legal, social, and critical theory. But, at the same time, it offers productive directions in environmental humanities, particularly through studies of posthumanism.

The 'spatial turn' occurred within geography and philosophy in the 1970s and 1980s, but it was not developed in the humanities until the early 2000s. Although not as recognised in literary and cultural scholarship, writers and artists have also employed critical spatial perspectives that respond to larger social justice movements both overtly and inadvertently (Williams 1973; hooks 1990; Jameson 2005 [1991]). Admittedly, the amount of 'turns' or qualifying titles have become dizzying and often repackage previous ideas more than forge completely new directions. And yet, they remain useful to create language that specifies theoretical and practical synergies and continuities between disciplines. Case in point, the geohumanities have arisen out of the need to understand the humanities from a geographical (and spatial) perspective. Authors of the book *GeoHumanities* define it as 'the rapidly growing zone of creative interaction between geography and the humanities' (Richardson et al. 2011, 3). By employing both spatial and environmental methodologies through the humanities, *Ecological Exile* could certainly be considered part of the geohumanities, as could any of the offerings mentioned in this section.

Spatial literary studies have also developed over the past several years, but the dearth of scholarship remains noticeable compared to ample amounts in geography. Robert T. Tally Jr.'s introductory book *Spatiality* (2013) is one notable guide to the 'spatial turn' in literary studies. Even though the specific relationship between space and justice is not necessarily the focus, the book examines many of the spatial thinkers who developed what we would now call spatial justice.[3]

Another recent and notable theory called geocriticism offers perhaps the most anticipated shift to spatial literary studies that captures links to social justice. In *Spatiality*, Tally maintains that geocritical scholarship should draw on theorists of geographical and social theory – Lefebvre, Foucault, Harvey, Massey, and Soja, among others such as Gaston Bachelard, Michel de Certeau, Gillian Rose, Gilles

Deleuze, and Félix Guattari – not only in critical theory. Combining social and cultural theory produces an understanding of both aesthetics and politics, which Tally argues provide 'a constellation of interdisciplinary methods designed to gain a comprehensive and nuanced understanding of the every-changing spatial relations that determine our current, postmodern, world'. Geocriticism profitably examines and uncovers 'hidden relations of power' in literary and cultural studies (Tally 2013, 113–14). What these examples reveal is that the relationship between space and place in literary and cultural studies is significant and currently on the rise (see more about place and geocriticism in the next section).[4] At the same time, they also show other necessary areas of exploration; scholars might investigate particular examples of space and social process in other forms of creative practices attentive to environmental issues not solely confined to literary studies.

To conclude this section, I will discuss two recent studies that, although not defined as such, I would argue are illustrative guides to spatial justice from the perspective of the environmental humanities (also focused on various forms of visual, literary, and audio texts). Timothy Morton's *Hyperobjects* (2013) offers another way to perceive the environmental crisis, or 'the end of the world' as he calls it, through a spatio-temporal lens. He defines real and imagined concepts such as climate change or oil (among many others) as 'hyperobjects', which are 'massively distributed objects in time and space' relative to human scale. They are 'hyper' in relation to some other entity, whether humans create these objects or not (Morton 2013, 1). Thinking about climate and oil spatially also pushes people out of the realm of object consciousness and into one about space, where perception and imagination supplements the real. Such hyperobjects are spatial as well as social constructs, largely produced by society as both a conceptual and physical reality. These might include uranium, the Pacific garbage gyre, climate, capital, or oil, all of which function in some way to prop up our current carbon-based world. Hyperobjects offer another way of understanding the immense catastrophe of something such as climate change without losing the immediacy of the local impacts on place and home.

For Morton, hyperobjects vary and are multitudinous. Drawing on a relatively new school of philosophical thought called object-oriented ontology (OOO), which was initially conceptualised by Graham Harman in 1999 and renamed OOO by Levi Bryant in 2009, Morton's analysis focuses on 'things' or objects, which challenge anthropocentric thinking by repositioning nonhuman objects. OOO serves Morton's ecologically focused approach because it not only removes humans from a privileged position, but it also provides space to understand an entire epistemology and ontology beyond the human. Put simply, objects exist outside of human perception; they contain their own ontology separate from humans. Humans and nonhumans exist as ontological equals. Such a position would support ecological or 'biocentric' thinking, which negates hierarchical status of the human in biological relations. This is why the highly spatial concept of hyperobjects, while seemingly theoretical, illuminates the unjust effects of changing climates produced by humans – by first differentiating the idea of climate from the changes it creates through weather patterns that ultimately affect the entire Earth.

The other study is Rob Nixon's roundly praised book *Slow Violence and the Environmentalism of the Poor* (2011). In it, Nixon outlines a different and slower way to understand ecological degradation as it is represented in the arts and humanities. While seemingly temporal, 'slow violence' contains spatial effects of environmental injustice as well because of the dispersal of time across vast spaces and scales. He argues:

> [slow violence] is a violence that occurs gradually and out of sight, a violence of delayed destruction that is dispersed across time and space, an attritional violence that is typically not viewed as violence at all. Violence is customarily conceived as an event or action that is immediate in time, explosive and spectacular in space, and as erupting into instant sensational visibility. We need, I believe, to engage a different kind of violence, a violence that is neither spectacular not instantaneous, but rather incremental and accretive, its calamitous repercussions playing out across a range of temporal scales. In so doing, we also need to engage the representational, narrative, and strategic challenges posed by relative invisibility of slow violence. (Nixon 2011, 2)

Violence for Nixon moves beyond the immediate spectacle and drama often associated with conventional forms of war and creates the long-term consequences of environmental abuses. This type of diffuse and dispersed violence, such as post-war landmines or toxicity in drinking water from industrial waste, generates postponed effects gradually experienced over time and space. Both humans and nonhumans suffer the effects of this violence as much as they might civil wars or political oppression. Dumping toxic waste in underdeveloped countries because no regulations exist can be just as violent as invading a country with military force. For Nixon, this 'slow-moving violence' is in many ways even more potent than traditional notions of violence; it is both 'attritional' and 'exponential'.

Nixon ultimately asks how writers and artists address these issues and then infuse them with some urgency. How do writers use, for example, narrative to address slow violence? The challenge is representational: how can we 'devise arresting stories, images, and symbols adequate to the pervasive but elusive violence of delayed effects?' (Nixon 2011, 3). While not acknowledged as such, Nixon's notion of slow violence reveals another form of spatial injustice, particularly in how it demonstrates time-space expansion (not capitalist compression), elongating the trauma of displaced violence over time and across various geographies. *Slow Violence* does, however, briefly underscore links to space and place through the works of Soja, Berger, and Said, among others.

As a postcolonial and transnational scholar, Nixon addresses various issues of displacement around the globe, which, as discussed in the next chapter, remains an important element of spatial injustice and forms of solastalgia. Slow violence, much like the complexity of Morton's hyperobjects, offers another example of cultural scholarship already concerned with the multiple ontologies used to confront and reproduce environmental injustices. Morton and Nixon's work finalise the critical discussion about spatial injustice because they both emphasise issues of environmental injustice through forms of spatial theory. More specifically, they demonstrate such approaches by examining specific instances represented in literature, film, media, music, and non-fiction, all of which emerge from artistic production in the humanities.

Place justice

This opening chapter has thus far explored some of the critical contours of spatial justice both broadly and specifically to the objectives of this book. Until now, there has been little discussion on the correlation between space and place, the latter of which underlies a fundamental element of solastalgia. Without oversimplifying too much here, place could be understood broadly as space imbued with meaning and experience within systems of hegemonic power (Cresswell 2015, 17–19). Place remains crucial to artistic/creative production because of its imaginative links to human perception and experience. To this effect, place continues to be a fundamental link between spatial justice and the environmental humanities.

Calling it the 'topographical turn' in geography and philosophy, Jeff Malpas, who is a philosopher of place studies, maintains that 'place is perhaps the key term for interdisciplinary research in the arts, humanities and social science in the twenty-first century' (2010). This is one reason why the notion of 'place' (*topos*) serves as a vital link between spatial and ecological approaches to the humanities; the focus on social and cultural formations of topographies and topologies – drawing out the heterogeneity of place – has produced expansive approaches from diverse disciplinary theory and practice.

Tim Cresswell's *Place* (2015) argues that place is as much a physical space as it is a way of perceiving, seeing, or understanding the world, particularly through humanistic epistemologies and ontologies of knowing and being. Understanding the many nuances of place opens comprehensions of meaning and experience, even while place is both pervasive and complex (Cresswell 2015, 18). Cresswell classifies place in three dominant categories: descriptive, socially constructed, and phenomenological.

Using a descriptive approach to place, one would identify the 'distinctiveness' of specific places, an attitude traditionally adopted by regional geographers. For instance, someone taking this direction would study the topography of the Yorkshire Moors to understand the distinctive aspects of this particular place not only in the UK, but also across the Earth. A social constructionist understanding would see the underlying social processes that shape and mould a particular place through meaning and materiality. This position emphasises vast inequalities that are caused by larger political and social structures of power. How are the social systems of capitalism or patriarchy fashioning structural conditions of a place and affecting people living in these conditions? A phenomenological idea of place does not focus on unique attributes or social forces constructing place, but instead it identifies the 'essence of human existence' as a necessary element of place. It explores humanistic and corporeal connections to place that accentuate experience

(Cresswell 2015, 56). Gaston Bachelard's book *The Poetics of Space* (1969) serves as a well-known example of this approach because it uses phenomenological approaches to investigate spaces of everyday life but specifically within the domestic realm. We could enquire how might a memory trigger connection to the real and imagined place in the past? Or, one could explore how might meaning be derived from a personal experience of place?

While these three approaches might appear somewhat contradictory, they do share commonality. To say, for example, that taking a social constructionist viewpoint of place (structural) does not also allow possible humanistic interpretations related to phenomenology (agency) would limit the multifarious ability of place to function as an expansive ontology and epistemology. When examining the underlying factors of climate change off the north coast of Wales, one might explore the cause – climate change is socially constructed through human endeavour – while also looking at the distinctiveness of a place and the essence of humans (and non-humans) in that place, such as the physical differences between the present and past and the memories and experiences forged from these landscapes that influence current and future behaviour. This circumstance is exemplified in Gillian Clarke's poem 'Cantre'r Gwaelod' discussed in the next chapter.

All three approaches to place seem complementary more than antagonistic, intersecting rather than separate. When looking at theories of spatial injustice as defined by Soja (drawing on Harvey, Foucault, Lefebvre, and others), they predominantly contain a social constructionist approach. Thus, they underlie the foundation of this book, which argues that because spaces (and places of solastalgia) are socially produced, they can also be reclaimed and socially reproduced in literary or visual narratives in the humanities (where artists have agency to shape new spaces) to combat forms of injustice related to oil and climate. In short, this is a form of place justice.

Taking a social constructionist approach to place aligns more broadly with social justice movements such as feminism, environmentalism, and Marxism, all of which confront the exploitation of power over, for instance, gender, class, or environments in social spaces. The general claim, with which I agree, is that inequalities of space lead to macro and micro forms of oppression. Controlling narratives of place therefore produce spaces of in/justice. Harvey succinctly sums up the dynamic between the social production of place and power when he identifies 'those who command space can always control the politics of place even though, and this is a vital corollary, it takes control of some place to command space in the first instance' (1990, 234). Environmental movements, much like working-class and anti-racist activism, excel at organising and controlling narratives around place more than they do at commanding and changing place. The imagined spaces of perception produced by various stories from an array of voices and groups are as important as (if not more important than) the real places in which people inhabit. This is ultimately why artists who engage with place justice in various ways attempt to reclaim and reconstruct place as a form of resistance through print, visual, or audio texts. A social constructionist approach to places of injustice underscores the ways in which society produces and responds to corrosive structures and systems that support injustices, such as petrochemical advocates who deny carbon-based energy systems contribute to global climate disasters.

No universal agreement over the definition of place exists, and yet it remains a contested concept that differs depending upon the usage and or disciplinary application. Ecologists and environmental activists identify place – often termed bioregions, which are geographical or regional areas larger than an ecosystem demonstrating patters of biodiversity – as pivotal to understanding ecological concerns. Place is also central to the work of urban planners and architects, where the historical and social implications of urban centres inform the spatial. Law relies on place to understand legal frameworks that no longer protect people, such as disenfranchised voting systems (gerrymandering) or instances of civil forfeiture. Literature has traditionally identified place (i.e. sense of place) as a setting or location for an imaginative novel or poem. D.H. Lawrence discussed a 'spirit of place', where a 'spirit' or essence of a place informs people who live there and the people who read about such places in fiction (1961, 6). Computer programmers and geographic information system (GIS) analysts now explore uses of virtual place, and the contestation over who controls democratised forms of virtual space. Regardless of its many critical applications, place continues to be conceptualised by geographers more than any other discipline, and so once again I will turn here to finish outlining critical approaches to place that fundamentally link to the subjects of the next chapter: solastalgia and the environmental humanities.

Yi-Fu Tuan and Edward Relph are two geographers who initially theorised some of the differences between space and place in the 1970s. Despite the passing of over 40 years, their work has constructively endured as foundational for contemporary approaches to place, and especially in humanistic disciplines drawing on real and imagined contexts. In *Topophilia* (1974), Tuan assessed humans' sensorial relationship to space and place through the concept of 'topophilia', literally meaning 'love of place'. For Tuan, there is 'the affective bond between people and place or setting' (1974, 136).

Elsewhere, Tuan explains how place is contingent upon subjective modes of experiential perspectives, such as sensation, perception, and conception though emotions and cognition, that undergo alteration depending upon each geography in which these phenomenological aspects are situated (1977, 3, 8). One of the ways space becomes place for Tuan is when a momentary 'pause' occurs in our larger 'view' of a space, an impressionable moment that imbues space with meaning (1977, 161). If space is movement, then place creates moments of pause. Tuan argues, 'From security and stability of place we are aware of the openness, freedom, and threat of space, and vice versa' (1977, 6). Embedded in this definition exists the 'threat' of space serving as a place of struggle, in addition to a place of openness and inclusivity.

Writers and artists often use topophilic approaches – exploring affective ties with material environments through characters, senses, values, perception, and narrative – developing the polymorphous nature of human experience connected to their environments. Tuan briefly references Wallace Stevens' famous poem 'Anecdote of the Jar' (1919) to illustrate how place becomes embedded in space materially as well as through our imagination. For Tuan, even though a geographer, art constructs the spaces around it. Stevens' poem not only resonates here because of its exhibition of place, but also because of its subtle nod toward poetry as a vehicle for understanding the complexities of place. Both reasons warrant extending the analysis of the poem further than Tuan.

> I placed a jar in Tennessee,
> And round it was, upon a hill.
> It made the slovenly wilderness
> Surround that hill.
>
> The wilderness rose up to it,
> And sprawled around, no longer wild.
> The jar was round upon the ground
> And tall and of a port in air.
>
> It took dominion everywhere.
> The jar was gray and bare.
> It did not give of bird or bush,
> Like nothing else in Tennessee.

The jar itself produces a place, where the 'wilderness' engulfs it to become place. The human-made material object of 'the jar', consisting of compounds of sand, limestone, and soda ash, necessitates fossil fuels for mass production. The jar is also a product of consumption, signalling its connection to capitalism and the economy of waste.

While we could read this poem as human 'dominion' over 'nature', a common interpretation among environmental scholars, we could also accentuate the social constructionist elements of place that underscores adaptations of ecology. The jar reconstructs the place of wilderness by challenging the perception of what we might consider the dominant idea of 'wilderness' to represent – which is to say, pristine and untouched wildness for human use. The problem here is that wilderness, similar to notions of nature and space, is a socially constructed concept. Put another way, place is not natural. No place is untouched. The conception itself assumes a social relationship associated with touching or not. Human forces produce places and they can undo them (Cresswell 2015, 46).

Stevens' poem begins with the speaker using the personal pronoun 'I' to accentuate how the jar arrived at this place. The 'anecdote', which is an experiential short story or narrative, is a reconstruction of space formed here through ecologically aware poetry. The jar, wilderness, bird, or bush are intertwined and interconnected, and this includes the systems of power such as petrochemicals and capitalism originally structuring the jar and the eventual location of the jar by a consumer. Literature, as well as other artistic forms, contains the ability to refocus attention on a place, turning an object into a marker of change where the influence of perception becomes reality. As Tuan puts it, 'culture affects perception' (1977, 162). The knowledge of this interconnection in the poem also sends a warning: 'it took dominion everywhere'.

Adding to this conversation initiated by Tuan, the famed ecologist E.O. Wilson addresses our 'love of life' – or innate connection to other nonhuman living organisms – in his Pulitzer Prize winning book *Biophilia* (2003 [1984]). Wilson argues the more we understand other organisms, the more we will 'place a greater value on them, and on ourselves' (2003, 1). The notion of biophilia deepens the understanding of topophilia by addressing how people come to love the organisms in a place as much as the actual place itself. Tuan focuses on human relationships to place, whereas Wilson looks at the relationship between place and organisms (human and nonhuman). When returning to the jar in Stevens' poem, we can see how it becomes an organism embedded in the wilderness, or even an object, challenging our notions of what it means to be 'natural' or the social processes that made the jar. Assuming the jar is out of place reinforces the constructed concept that an untouched wilderness exists. The jar is ultimately at 'home', despite its material structure, because it is now part of the speaker's memory of the place that he observes over time.

Relph's comparison of space and place in his book *Place and Placelessness* (1976) is also useful to cite briefly here because it partly supports the concept of solastalgia. He argues, 'Space is amorphous and intangible and not an entity that can be directly described and analysed. Yet, however we feel or explain space, there is nearly always some associated sense or concept of place' (Relph 1976, 8). Similar to Tuan, Relph assigns space a role separate from place, abstracting space as a container compared to the experience of place. Relph goes on to suggest the 'essence of place lies in the largely unselfconscious intentionality that defines places as profound centers of human existence' (1976, 43).[5] Both Tuan and Relph offer humanistic approaches to place, drawing on phenomenology, but they also imply distinctiveness and social constructionist elements to place, as revealed in Stevens' poem 'Anecdote of the Jar'.

Tuan and Relph, in addition to Cresswell, all prove in comparative and contrasting ways that an essential relationship exists among place, organisms (including humans and nonhumans), and ecology (interconnectedness of organisms). Tuan and Relph may appear outdated, but they still resonate – particularly in the environmental humanities where analyses of place as an ecological concept are not as apparent. Moving into the twenty-first century, amid globalisation and lives of constant mobility largely supported by an equally transportable capitalist world-system, place remains an important concept to understand and deploy when confronting forms of spatial injustice. One promising area of development for the environmental humanities and place justice is in the field of geocriticism.

In the definitive book outlining geocritical theory entitled *Geocriticism: Real and Fictional Spaces* (2007; trans. 2011), French literary critic Bertrand Westphal argues that geocriticism 'probes the human spaces that the mimetic arts arrange through, and in, texts, the image, and cultural interactions related to them' (2011, 6).

The primary concern in Westphal's geocriticism is the interdisciplinary study of place; the cultural production surrounding texts remain secondary for the geocritic. A geocritical approach would analyse a text by first spotlighting its specific place – geographic location, landscape, bioregion, or environment – and then its relationship to a writer, text, or political context. For Westphal, human spaces, viewed as a conflation of totalising space and personal place, mirror and inform our relationship to them.

Westphal goes on to suggest that geocriticism situates the complex notion of place at the centre of spatial studies, whereas other approaches to space within similar literary or cultural theories of the environment do not, such as imagology, ecocriticism, or geopoetics (2011, 112). Geocriticism differs from these other theories because it avoids the 'egocentric logic' that interprets how literature or culture informs a place through an individual perspective (author or character). 'Geocentric logic' interrogates how a real place informs the writer, work, and audience, which would more directly apply to my approach in this book (Westphal 2011, 111). Geocentric logic decentres the human as a primary focus, while also merges with the aims of spatial justice in the ways they both address transformative effects of postcolonialism, globalisation, deterritorialisation, and environmentalism.

Simply put, Westphal's geocriticism is a study of space that explores 'real' places and their relationship to fictional or 'imagined' places (echoing Lefebvre's experienced and perceived spaces). The converse is also true: fiction allows us to understand these real places in multidimensional ways (Tally 2011, x). For example, instead of focusing on James Joyce's representation of early twentieth-century Dublin in works such as *Ulysses* (1922) or *Dubliners* (1914), a geocritical approach would first analyse the real twentieth-century Dublin, called the 'geographical referent', and then investigate its relationship to the fictional or imagined world represented by Joyce in the literary text. The interplay between real and the fictional representations of a place, what Westphal calls 'the interface of the world and text', furthers our understanding of how space functions in not only literary and cultural texts, but, more importantly, in the physical world (2011, 112). Real instances of injustices related to space, such as forms of colonisation or displacement of the urban poor, could be analysed alongside a literary depiction describing these examples. Although the progenitor of the theory, approaches to geocriticism extend beyond Westphal and would complement work in the geohumanities.[6]

Notwithstanding the volumes written about space and place in critical geography in the years between Tuan and Cresswell, as well as more recent work in geocriticism and the geohumanities, approaches to place and space remain more peripheral in the environmental humanities, and particularly in literary and visual disciplines.[7] In *The Future of Environmental Criticism* (2005), Lawrence Buell acknowledges literary and cultural studies focused on environmental issues need to evolve and embrace the spatial turn. Buell suggests that one of the ways to achieve this aim is by examining the notion of place – what he defines as the 'quotidian idiosyncratic intimacies' that emerge from more totalising and sometimes abstract forms of space (2005, 63). The focus on place furthers Buell's understanding of ecology and serves as a key point of intersection between spatial and environmental studies. According to Buell:

> Place is an indispensable concept for environmental humanists not so much because they have precisely defined and stabilized it as because they have not; not because of what the concept lays to rest as because of what it opens up. (2005, 62)

Buell's invitation to think more spatially in the realm of the environmental humanities reinforces the necessity to theorise space more assiduously because the concept of place remains 'an additionally rich and tangled arena for environmental criticism' (2005, 63).

Spatiality 'opens up' our understanding of the environmental humanities by revealing the complexities of real and representational places, particularly in literary, visual, and cultural texts. Various environments, as real (experienced) and imagined (conceived) spaces, become repositories of personal and cultural memory through the practice of what Buell calls 'place-attachment' (2005, 63). In other words, place carries meaning because it reflects an emotional, historical, and cultural attachment for people to an existing space. Not only is place 'associatively thick' and 'defined by physical markers as well as social consensus'; it also demonstrates a collective chronicle where the histories of space become place through the 'quotidian idiosyncratic intimacies that go with "place"' (Buell 2005, 63).

In one final example, Ursula Heise examines place in her book *Sense of Place and Sense of Planet* (2008) as a form of ecolocalism in literature and film where the global is imagined through the frame of the local. Rather than examine place as a critical framework, it is used to distinctively mark ways of perceiving the local through the lens of a global eco-cosmopolitism. Heise warns that 'an individuals' sense of place' ultimately 'becomes a visionary dead end' if such place-thinking leads to reductive views of 'nature', instead of understanding the wider local-global nexus. Heise argues:

> Rather than focusing on the recuperation of a sense of place, environmentalism needs to foster an understanding of how a wide variety of both natural and cultural places and processes are connected and shape each other around the world, and how human impact affects and changes this connectedness. (2008, 8–9, 21)

Heise's warning takes us immediately back to Stevens' 'Anecdote of a Jar' as a literary text that fosters both a sense of place and a critique of how socially constructed systems of capitalism and consumption shape and impact the 'natural and cultural places and processes' without leading to reductive views of 'nature'.

Heise's caution might additionally refer to other utopian visions of place represented in literary or film culture, which are commonly conceptualised in the North American popular environmental imagination. However, work in spatial and place justice outside of literary and visual scholarship often focuses on the cultural and environmental processes that shape each other globally. For example, Buell's idea of place-attachment or Albrecht's notion of solastalgia examined in the next chapter underscore the meaning of place for both the local and global community, demonstrating that interconnected social formations affect all parts of the global network. Experiencing and losing place-attachment is also what often creates pain or sickness when spatial injustices occur, conjuring a feeling of displacement – or loss of place related to solastalgia – while remaining in the same geography. It is this simultaneous attachment to place, even while many places are disappearing around the globe as a result of spatial injustices and thereby affecting environments where people live, that is causing populations to suffer from solastalgia.

Spatial justice is ultimately a way of investigating some of the unjust circumstances – both influenced by and separate from society and history – fundamentally affecting humans and nonhumans around the globe. It also illuminates new ways of thinking about and provoking change in the places we live (Soja 2010, 5). The contemporary spatial turn looks more at collective living on a planetary scale during a time in the twenty-first century when energy resources are more limited and where increased global temperatures are at critical levels – more than temperatures of the Earth were several decades ago in the 1960s (Conley 2012, 1). I shall move beyond the sole emphasis of urban injustices as places of social and spatial discontinuity, as in the case of most spatial theorists, and instead examine how we might understand more contemporary environmental concerns related to fossil fuels and climate change. One of these areas of further research is the unanswered question about the ways in which artistic production in the environmental humanities respond to, confront, or affirm forms of spatial injustices. Because humans are inherently spatial beings, and creative works draw out these elements, space is important to any discipline that engages with society and culture. There are many issues that could be tackled about space and social processes that emerge in the humanities, but this book is focused particularly on the environmental or ecological circumstances of loss of place, as discussed in the next chapter as solastalgia, that result from spatial injustices as they are represented in some literary and visual culture.

Notes

1 The following list is not exhaustive, but it offers some of the influential works on the subject. See Harvey (1996a, 2003), Davis (1968), Lefebvre (1968, 1991), Coates et al. (1977), Foucault (1980, 1982, 1986), de Certeau (1984), Soja (1989, 1996, 2010), Rose (1993), Blomley (1994), Gregory (1994), Massey (1994), and Smith (2008).

2 While the original version of *La production de l'espace* was published in 1974, it was translated into English in 1991. Because of this publication history and availability in English, it remains a relatively recent approach to contemporary responses in theory and practice to spatial injustices.

3 Tally's forthcoming anthology is a welcomed addition to the growing field of literary spatial studies (see Tally 2017). Other works focused specifically on British and Irish spatial concerns related to literary and visual culture include Smyth (2001), Colombino (2013), Duff (2015), and Yeung (2015).

4 Tally's new edited book series with Palgrave Macmillan entitled *Geocriticism and Spatial Literary Studies* reveals the significant influx of spatial scholarship in the humanities. Since 2014, the series has published 13 books focused on the dynamic relations among space, place, and literature.
5 I am grateful to Cresswell's thorough introduction to Tuan and Relph in *Place* (Chapter 2), which allowed me to pursue these leads further.
6 Tally challenges Westphal's somewhat 'narrow geocentric approach'. For Tally, the spatial turn in literary and cultural studies opens up 'new spaces for critical inquiry, in which we may see the ways that wirers map their world and readers engage with such literary maps' (2013, 144). Whereas Westphal prescribes a specific approach to geocriticism, Tally sees it as a broader way of categorising space, place, and mapping in literature and culture.
7 Two notable exceptions to this include Heise (2008) and Tally and Battista (2016).

Bibliography

Bachelard, Gaston. 1969 [1964]. *The Poetics of Space: The Classic Look at How We Experience Intimate Places*. Translated by Maria Jolas. Boston, MA: Beacon Press.

Blomley, Nicholas. 1994. *Law, Space, and the Geography of Power*. New York: Guilford Press.

Bromberg, Ava, Gregory Morrow, and Deirdre Pfeiffer. 2007. 'Editorial Note: Why Spatial Justice?' *Critical Planning* 14: 1–6.

Buell, Lawrence. 2005. *The Future of Environmental Criticism: Environmental Crisis and Literary Imagination*. Cambridge, MA: Blackwell.

Coates, Bryan E., R.J. Johnston, and Paul C. Knox. 1977. *Geography and Inequality*. Oxford: Oxford University Press.

Colombino, Laura. 2013. *Spatial Politics in Contemporary London Literature: Writing Architecture and the Body*. London: Routledge.

Conley, Verena Andermatt. 2012. *Spatial Ecologies: Urban Sites, State and World-Space in French Cultural Theory*. Liverpool: University of Liverpool Press.

Cresswell, Tim. 2015 [2004]. *Place: An Introduction*. Malden, MA: Wiley Blackwell.

Davis, Bleddyn. 1968. *Social Needs and Resources in Local Services: A Study of Variations in Provision of Social Services between Local Authority Areas*. London: Joseph Rowntree.

Dear, Michael. 2011. 'Creative Places: Geocreativity'. In *GeoHumanities: Art, History, Text at the Edge of Place*, edited by Michael Dear, Jim Ketchum, Sarah Luria, and Douglas Richardson, 6–11. New York: Routledge.

De Certeau, Michel. 1984. *The Practice of Everyday Life*. Translated by Steven Rendall. Berkeley, CA: University of California Press.

Duff, Kim. 2015. *Contemporary British Literature and Urban Space: After Thatcher*. Basingstoke, UK: Palgrave Macmillan.

Duncan, James S., Nuala C. Johnson, and Richard H. Schein. 2004. 'Introduction'. In *A Companion to Cultural Geography*, edited by James S. Duncan, Nuala C. Johnson, and Richard H. Schein, 1–8. Oxford: Blackwell.

Flusty, Steven. 1994. *Building Paranoia: The Proliferation of Interdictory Space and the Erosion of Spatial Justice*. West Hollywood, CA: Lost Angeles Forum for Architecture and Urban Design.

Foucault, Michel. 1980. 'The Eye of Power'. In *Power/Knowledge: Selected Interviews and Other Writings, 1972–1977*, edited by Colin Gordon, 146–65. New York: Pantheon.

——. 1982. 'Space, Knowledge, and Power'. In *The Foucault Reader*, edited by Paul Rabinow, 239–56. New York: Pantheon.
——. 1986. 'Of Other Spaces'. Translated by Jay Miskowiec. *Diacritics* 16: 22–7.
——. 2007. 'Questions on Geography'. In *Space, Knowledge and Power: Foucault and Geography*, edited by Stuart Elden and Jeremy W. Crampton, translated by Colin Gordon, 173–82. Hampshire, UK: Ashgate.
Gregory, Derek. 1994. *Geographical Imaginations*. Oxford: Blackwell.
Harvey, David. 1990. *The Condition of Postmodernity: An Enquiry into the Origins of Cultural Change*. Oxford: Blackwell.
——. 1991. Afterword to *The Production of Space*, by Henri Lefebvre, 425–34. Oxford: Blackwell.
——. 1996a. *Justice, Nature, and the Geography of Difference*. Oxford: Blackwell.
——. 1996b. *Writing on Cities/Henri Lefebvre*. Oxford: Blackwell.
——. 2009 [1973]. *Social Justice and the City*. Athens, GA and London: University of Georgia Press.
Heise, Ursula K. 2008. *Sense of Place and Sense of Planet: The Environmental Imagination of the Global*. Oxford: Oxford University Press.
hooks, bell. 1990. *Yearning: Race, Gender, and Cultural Politics*. Boston, MA: South End Press.
Jameson, Fredric. 2005 [1991]. *Postmodernism: Or, the Cultural Logic of Late Capitalism*. Durham, NC: Duke University Press.
Joyce, James. 1986 [1922]. *Ulysses*. Edited by Hans Walter Gabler. New York: Vintage.
Joyce, James. 1992 [1914]. *Dubliners*. New York: Penguin.
Kearns, Gerry, David Meredith, and John Morrissey, eds. 2014. *Spatial Justice and the Irish Crisis*. Dublin: Royal Irish Society.
Lawrence, D.H. 1961. *Studies in Classic American Literature*. New York: Vintage.
Lefebvre, Henri. 1968. *Le Droit á la Ville*. Paris: Anthropos.
——. 1974. *La Production de l'espace*. Paris: Anthropos.
——. 1991 [1974]. *The Production of Space*. Translated by Donald Nicholson-Smith. Oxford: Blackwell.
Malpas, Jeff. 2010. 'Place Research Network'. *Progressive Geographies*, 4 November. Accessed 7 October 2016. https://progressivegeographies.com/2010/11/04/place-research-network/.
Massey, Doreen. 1994. *Space, Place, and Gender*. Minneapolis, MN: University of Minnesota Press.
Morrissey, John. 2014. 'Challenging the Political Economies of Injustice: An Interview with David Harvey'. In *Spatial Justice and the Irish Crisis*, edited by Gerry Kearns, David Meredith, and John Morrissey, 209–23. Dublin: Royal Irish Society.
Morton, Timothy. 2013. *Hyperobjects: Philosophy and Ecology after the End of the World*. Minneapolis, MN: University of Minnesota Press.
Nixon, Rob. 2011. *Slow Violence and the Environmentalism of the Poor*. Cambridge, MA: Harvard University Press.
O'Laughlin, John. 1973. *Spatial Justice for the Black American Voter: The Territorial Dimension of Urban Politics*, PhD diss., Department of Geography, Pennsylvania State University.
Philippopoulos-Mihalopoulos, Andreas. 2015. *Spatial Justice: Body, Lawscape, Atmosphere*. London: Routledge.
Pirie, Gordon H. 1983. 'On Spatial Justice'. *Environmental and Planning A* 15: 465–73.
Rawls, John. 1971. *A Theory of Justice*. Cambridge, MA: Harvard University Press.

Relph, Edward. 1976. *Place and Placelessness*. London: Pion.

Richardson, Douglas, Sarah Luria, Jim Ketchum, and Michael Dear. 2011. 'Introducing the Geohumanities'. In *GeoHumanities: Art, History, Text at the Edge of Place*, edited by Michael Dear, Jim Ketchum, Sarah Luria, and Douglas Richardson, 3–5. New York: Routledge.

Rose, Gillian. 1993. *Feminism and Geography: The Limits of Geographical Knowledge*. Minneapolis, MN: University of Minnesota Press.

Said, Edward. 1993. *Culture and Imperialism*. New York: Vintage.

Scott, Heidi. 2004. 'Cultural Turns'. In *A Companion to Cultural Geography*, edited by James S. Duncan, Nuala C. Johnson, and Richard H. Schein, 24–37. Oxford: Blackwell.

Smith, David M. 1994. *Geography and Social Justice*. Oxford: Blackwell.

Smith, Neil. 2008 [1984]. *Uneven Development: Nature, Capital, and the Production of Space*. Athens, GA and London: University of Georgia Press.

Smyth, Gerry. 2001. *Space and the Irish Cultural Imagination*. Basingstoke, UK: Palgrave.

Soja, Edward. 1989. *Postmodern Geographies: The Reassertion of Space in Critical Social Theory*. London: Verso.

——. 1996. *Thirdspace: Journeys to Los Angeles and Other Real-and-Imagined Places*. Malden, MA: Blackwell.

——. 2009. 'The City and Spatial Justice / La ville et la justice spatiale'. Interview with Sophie Didier and Frédéric Dufaux, *Spatial Justice / Justice spatiale* 1: 1–5.

——. 2010. *Seeking Spatial Justice*. Minneapolis, MN: University of Minnesota Press.

——. 2013. 'Seeking Spatial Justice in Asian Cities'. Lecture, Lee Kuan Yew School of Public Policy, Singapore, 23 January. Accessed 3 July 2014. www.youtube.com/watch?v=I7Blgo4uY9s.

Stevens, Wallace. 1919. 'Anecdote of the Jar'. The Poetry Foundation. Accessed 12 December 2016. www.poetryfoundation.org/poetrymagazine/poems/detail/14575.

Tally Jr., Robert T. 2011. 'The Timely Emergence of Geocriticism'. Translator's preface to *Geocriticism: Real and Fictional Spaces*, by Bertrand Westphal, ix–xiii. London: Palgrave Macmillan.

——. 2013. *Spatiality*. New York: Routledge.

——. 2017. *The Routledge Handbook of Literature and Space*. New York: Routledge.

——, and Christine M. Battista, eds. 2016. *Ecocriticism and Geocriticism: Overlapping Territories in Environmental and Spatial Literary Studies*. New York: Palgrave.

Tuan, Yi-Fu. 1974. *Topophilia: A Study of Environmental Perception, Attitudes, and Values*. New York: Columbia University Press.

——. 1977. *Space and Place: The Perspective of Experience*. Minneapolis, MN: University of Minnesota Press.

Yeung, Heather H. 2015. *Spatial Engagement of Poetry*. New York: Palgrave Macmillan.

Young, Iris Marion. 2011 [1990]. *Justice and the Politics of Difference*. Princeton, NJ: Princeton University Press.

Westphal, Bertrand. 2007. *La Géocritique: Réel, Fiction, Espace*. Paris: Les Éditions de Minuit.

——. 2011. *Geocriticism: Real and Fictional Spaces*. Translated by Robert T. Tally Jr. New York: Palgrave Macmillan.

Williams, Raymond. 1973. *The Country and the City*. Oxford: Oxford University Press.

Wilson, E.O. 2003 [1984]. *Biophilia: The Human Bond with Other Species*. Cambridge, MA: Harvard University Press.

2 Solastalgia and the environmental humanities

One primary task of this book is to explore how we align concepts that shape spatial justice and place with the environmental humanities. In doing so, this chapter builds upon the previous chapter in Part I by critically overviewing two essential concepts that foreground the following two parts of the book. This chapter explains the links between place justice and solastalgia as part of the environmental humanities. It explores how socially produced ideas about space and environments in literary and visual texts affect the way we perceive and react to our environments. Establishing and defining a connection to place (locally, regionally, and globally) is what ultimately promotes change through spatial awareness.

Initially, I will examine the process of solastalgia as a way of understanding place-based environmental tensions. This overview sets up the next section, which provides a more extensive explanation of the environmental humanities, along with an applied analysis of a Greenpeace UK viral YouTube video by the London-based agency Don't Panic that confronts drilling in the Arctic entitled *LEGO: Everything Is NOT Awesome* (2014), along with a shorter analysis of the poem 'Cantre'r Gwaelod' (2015) by the Welsh poet Gillian Clarke. Finally, to show instances of solastalgia caused by spatial injustice and why this is a concern for the environmental humanities before moving into the following two parts on oil and climate, the last section of the chapter investigates Eavan Boland's poem 'In Our Own Country' (2007), which is paired with Oliver Comerford's short film *Distance* (2004) as a unified filmpoem available on the open-access website The Poetry Project: Poetry and Art from Ireland. Here, I argue that Boland and Comerford through both visual and print texts perceive social layers of place, with accumulated pasts, histories, and memories that all underscore place-home through the geocritical notion of stratigraphic vision. These creative works exemplify the phenomenon of solastalgia as represented in literary and visual texts within the environmental humanities.

Displacement of place-home

When spatial injustices occur, one of the major effects is that they geographically, as well as emotionally and psychologically, displace human and nonhuman populations. The notion of displacement, where people are physically forced to leave their homes, might seem incongruent with what I am paradoxically calling 'ecological exile'. When humans and nonhumans suffer the environmental consequences created by external forces, but have no choice but to remain in their home environment, it creates another form of dislocation or ecological exile called 'solastalgia'.

In December 2015, Storm Desmond overpowered Lancaster in the UK, cutting off power to 55,000 residents and displacing many others for weeks. Once the floods subsided, residents returned to the flood-ravaged areas in which they had always lived, but now with an altered environment around them. The burst of flooding between 2014 and 2016 is partly caused because of the systematic removal of nearby wetlands for urban development. The other cause is because of increasing global temperatures and subsequent extreme weather. Researchers at the University of Salford concluded that the severity of the flooding results from a breakdown of rivers and floodplains. Floodplains should adequately provide catchments for excess water from heavy rainfall. Unfortunately, rivers in this area no longer have this capacity. Scientists warn that with continual losses to soil and nutrients in agricultural land, flooding will increase 9 per cent by 2050 ('Drone' 2016).

The example of the Lancaster floods serves as one of many instances not only in the UK, but also around the globe (e.g., the recent catastrophic flooding in Texas and in Bangladesh, Nepal, and India), where environmental destruction caused by preventable actions displace populations. The fallout from the flooding still affects residents of Lancaster, with many businesses having to close because of flood damage and families who have lost their homes. One resident, Rohina Caterina, observed that 'this is what it must have been like in the war' (Pidd 2015). Environmental consequences created by unsustainable development practices amid an ever-changing global weather system are rapidly rising. These effects create new forms of place-based distresses for those people affected and illuminate the issue of solastalgia.

In *Getting Back into Place*, Edward Casey explains how 'place pathology' relates to issues of nostalgia, specifically in the context of Indigenous populations and Western culture. He argues, 'Nostalgia, contrary to what we usually imagine, is not merely a matter of regret for lost times; it is also a pining for lost places, for places we have once been in yet can no longer reenter' (Casey 1993, 37–8). Casey points out that while non-Indigenous people around the planet have lost their place, and also experience the outcomes of environmental injustices, Indigenous people have suffered doubly so by losing their sense of place and their land (1993). For instance, there is a difference between witnessing developers draining wetlands to build condominiums across the street from a middle-class neighbourhood and experiencing a mining company acquire your land to extract minerals from it, all while you still live on it. What were once more traditional forms of colonisation have become neocolonisation, where people are dispossessed without being forcibly removed, but are left to suffer the health and psychological effects of environmental, cultural, and psychic pollution. Systematic conditions where people experience loss of land and loss of place have historically produced tangible conditions of distress from displacement, such as homelessness, depression, and memory loss. These examples can all be symptoms of solastalgia.

Rather than examining lost places related to nostalgia, which is experienced when people are forced to leave where they live, solastalgia demonstrates contemporary instances of displaced people still living in places that remain, but which are significantly transformed. In these instances, people experience place-based grief connected to powerlessness when faced with losing their home to environmental injustices. People still live 'at home', but are confronted with similar feelings of being displaced as those historically suffering from nostalgia, where there is a severed relationship between their psychic identity and home. Any solace or comfort associated with 'place' and 'home' have been lost (Albrecht 2005, 44).

Glenn Albrecht is a philosopher of environmental ethics and sustainability who champions transdisciplinary research to unify intellectual frameworks beyond singular disciplinary perspectives. Applying his work on solastalgia to the arts and humanities seems to be both apt and welcomed. Albrecht's research broadly examines the interplay between ecosystem damage and human distress. He originally described this condition as 'ecosystem distress syndrome', but later developed the concept of solastalgia where it was first introduced at the Ecohealth Conference in Montreal, May 2003 (Albrecht 2005, 41–2). As a way to explain this unique condition, Albrecht draws from concepts of 'solace and 'desolation', both of which derive from *solari* and *solacium*, and attempt to alleviate distress during traumatic events or experiences. Solastalgia can literally be defined as 'the pain or sickness caused by the loss or lack of solace and the sense of isolation connected to the present state of one's home and territory' (Albrecht 2005, 45).

The term nostalgia etymologically derives from the combination of *nostos* (return home or native land) and *algia* (pain or sickness). It provides a language to describe the experience of how people, who have been dispossessed of their land and cultural meaning, desire to be back 'home'. In contrast, solastalgia stems from *solari* (relief) and *algia* (pain or sickness). It describes the feeling of being exiled in one's home or place without ever completely leaving what I am calling 'place-home' as a link between the two. Despite being at home, one continues to suffer distress and melancholia from experiencing ecological damage to place (Albrecht 2006, 34–5). Nostalgia explains the circumstances when a person lives in two separate geographical places simultaneously, while only physically inhabiting one of those spaces. This stressful effect disturbs one's sense of place. Solastalgia describes the condition in which a person lives in *one* geographical location, but with the lived experience of desolation and dislocation, despite still being at home, due to how that geographical space has been altered through environmental degradation. Had people left Lancaster because of the flooding, they might have suffered from nostalgia, but having to return to altered conditions because of severe water damage could trigger feelings of solastalgia, affecting people emotionally, psychologically, or physically.

The concept of solastalgia provides another way to describe the symbiotic connection between ecosystem and human health through the concept of place as home. As mentioned in the introduction of this book, the linguistic origins of ecology stem from the Greek word *oikos* meaning 'house' or 'environment', conjuring similar notions of place and home. One's connection to place-home indicates a generative place to live, rooted in identity, culture, and family. For Albrecht, environmental damage causes a 'homesickness' related to the feeling of solastalgia because environmental injustice generates a 'place-based distress' resulting in feelings of powerlessness (2005, 44). Desolation conjures similar responses as solastalgia might, but solastalgia induces feelings of abandonment and isolation, particularly in relation to something such as 'devastated land' (Albrecht 2006, 35). Albrecht argues:

> Dispossession is one trigger, but environmentally induced distress also occurs in people who are not displaced. There are places on Earth that are not completely lost, but are radically transformed. People who are neither voluntarily nor forcibly removed from their homes can experience place-based distress in the face of the profound environmental change. (2006, 35)

Albrecht surmises that 'environmentally induced stress' largely causes solastalgia. This lived experience that populations suffer from explains how they 'lack the sense of solace or comfort that is usually derived from their relationship to home' (Albrecht 2006, 35).

Solastalgia, as well as nostalgia, is a psychoterratic illness experienced in particular instances of dispossession and displacement. As Albrecht and his colleagues explain, 'Psychoterratic illness is defined as an earth-related mental illness where people's mental wellbeing (psyche) is threatened by the severing of "healthy" links between themselves and their home/territory' (Albrecht et al. 2007, 95). Often caused by war, terrorism, colonisation, or natural disasters, psychoterratic illnesses highlight the forced separation from home. Solastalgia specifically exists when environmentally caused stress is 'the lived experience of the physical desolation of home' (Albrecht et al. 2007, 96). The lack of solace or comfort that comes from a 'lived experience' of attachment to 'home' serves as a new type of psychoterratic illness that directly relates to ecological effects of one's place-home.

Albrecht and his colleagues apply an example of solastalgia to a specific case study of a large-scale open-cut coal mine and two large power stations (Bayswater and Liddell) in the Upper Hunter region of New South Wales (NSW) in Australia. From 1987 to 2004, there were profound environmental changes with long-term impacts on the health of the ecosystem. Documented illnesses from these projects, also known as 'somaterratic' (body-earth) illnesses affecting the physical body, are linked to toxic pollution and particulate fallout. During a three-year period from 2003 to 2006, a Newcastle-based research group interviewed community members living in the Upper Hunter. What became evident from the interviews is that everyone registered personal and emotional responses to the impacts of the mining and power stations. Albrecht and his research team concluded that the transformed topographies of the Upper Hunter directly caused solastalgia for the Indigenous community living there. As they explain, 'Their sense of place, their identity, physical and mental health and general wellbeing were all challenged by unwelcome change' (Albrecht et al. 2007, 96). The people interviewed also felt helpless to the unjust and systemic loss of place in their home environment.

As the Upper Hunter example validates, solastalgia can cause illnesses that are conceptual and empirical in scope. While these instances directly relate to deteriorating environmental conditions experienced by Indigenous people in Australia, they also parallel other disenfranchised populations in the Western world who feel loss of place because of urbanisation, soil erosion, wetland degradation, or drought and flooding (Albrecht 2005, 46, 49).[1] In fact, other more privileged populations feel the effects of solastalgia on some level through urban development or the consequences of fracking (hydraulic mining) in rural areas. The greater the global environmental problems, the more all populations will begin or continue to experience various forms of solastalgia.

Solastalgia, like nostalgia, is an effect and affect – or a way of describing an experience of distress in the places people live, often caused by another force. While functioning as a psychological and phenomenological explanation, or an *affect*, solastalgia also provides ways of understanding the *effects* of spatial injustice on people and places. What I want to shed light on is how this effect is confronted and narrated in the humanities, thereby acknowledging the loss of place-home – or the relationship between affective bonds between our sense of place in combination with an association of home. Although Albrecht works from an eco-psychological perspective, his approach to understanding and articulating the feeling associated with what I refer to as the 'displacement of place-home' as a paradoxical type of ecological exile is given expression in literature and visual culture.

For these reasons, I am drawing on the concept of solastalgia as another method to explore the bond between spatial justice and the environmental humanities. *Ecological Exile* underscores what Albrecht calls 'lived experience' (unintentionally echoing Lefebvre) that values place as home and acknowledges the impact of the physical desolation of peoples' homes or territories (2006, 35). The environmentally induced stresses causing solastalgia might seem a bit vague, creating distance from something as intimate as home or place, but it refers to both human-caused (colonialism, war, mining, land-clearing, fracking, or climate change) and organic events (natural disasters, such as flooding or earthquakes), which are increasingly produced by human systems and exacerbate climate change. The fallout caused by 'profound environmental change' is largely a spatial concern, as well as a historical, social, or cultural issue. People feel powerless about reproducing the social and spatial conditions to combat or improve their sense of desolation and displacement, particularly when state industries, in addition to multinational corporations, promote projects that cause ecological stresses.

Instances of solastalgia result from spatial injustices in specific places. For instance, we can think of the Standing Rock Sioux Territories water protection protests against the Dakota Access Pipeline (DAPL) in the United States as not only a place or location, but also as a way of understanding and knowing about a loss of place for specific Indigenous peoples. Do we continue to socially produce these places as 'wasteland' open for pipeline access? Or, can we reconstruct the narrative in the social order to highlight the anti-environmental, racist, and capitalist aims behind rerouting the DAPL through protected water sources?

On an even larger scale, solastalgia places Earth as a place-home. In a climate change poem by the Irish writer and poet Theo Dorgan entitled 'The Question', it concludes by asking a series of existential questions that transcend the material boundaries of Earth:

> What have you done
> With what was given you,
> what have you done with
> the blue, beautiful world? (2015)

The questions appear rhetorical but are obviously addressed to humanity. The microcosmic instances of solastalgia, as exemplified in the Upper Hunter or Standing Rock Sioux Territories, represent a macrocosmic issue of sustaining survival on the Earth. Dorgan's poem underscores prevalent metaphors of stewardship, reframed less anthropocentrically as responsibility, but it also goes deeper by asking one of the most important questions facing humans: what have you done? The consequences are universal, not only relegated to a few.

Solastalgia is a universal issue for everyone, though it largely affects those suffering from forms of spatial injustice. Because solastalgia is a problem chiefly induced by and affecting humans, it therefore remains an issue of injustice. Various examples of solastalgia are represented in literature, film, and media culture, all of which have responded in various ways to real and imagined situations that reflect some of the poignant examples of environmental damage as a consequence of space.

Humanities for the environment

A question that remains relatively unanswered in the environmental humanities is how society, particularly reflected in creative works, thinks not only temporally (historically), but also spatially about environmental justice. The environmental humanities were partly formed by historians and so temporal trajectories of the environmental crisis have tended to dictate critical engagement. My aim here is to add a spatial dimension to some of the groundbreaking work already conducted about time (e.g. deep time). After all, the Anthropocene is both temporal and spatial, measured by time and scale. As Timothy Morton posits, 'In order to have an environment, you have to have a space for it; in order to have an *idea* of an environment, you need ideas of space (and place)' (2007, 11; original emphasis). Multidisciplinary relationships among space, justice, and the environment in the humanities remain as vital as histories but need more attention.[2]

Geographers have, in some cases, extended the analysis of space to environmental justice (Harvey 1996; Soja 2009; Whitehead 2014). Although race, class, and gender constitute approaches to critical spatial studies, environmental justice is equally part of this grouping. As the environmental sociologist John Bellamy Foster reminds us:

> the crisis of the earth is not a crisis of *nature* but a crisis of *society*. The chief causes of the environmental destruction that faces us today are not biological, or the product of individual human choice. They are social and historical, rooted in the productive relations, technological imperative, and historically conditioned demographic trends that characterize the dominant social system. (1999, 12; original emphasis)

What Bellamy calls 'productive relations' encompass the spatial as much as the social and historical. Territorial inequalities in contemporary urban and rural geographies are both causes and effects of environmental crises throughout the planet – from population displacement and uneven urban development to environmental degradation, pollution, and climate change. If spaces and our perceptions of environments and 'nature' are socially produced, then they are changeable by challenging dominant social systems through collective perceptions and narratives about the state of the world. A powerful way this process occurs is through our exposure to some forms of artistic and creative production accessible in our daily lives.

If compelled to do so, artistic producers in the humanities are often the first to respond to and confront pertinent social circumstances in complex and sophisticated ways, whether it is through creative forms of literature, film, photography, art, or multimedia. When reading a novel or watching a film about the effects of climate change, we are introduced to sympathetic characters in relatable and intimate settings. If we invest in these characters' stories through narrative, used as a storytelling of resistance because they often reflect reality, we are more likely to perceive how people and places are disturbed by droughts or flooding than we might be by watching the news or reading scientific articles in peer-reviewed journals. The emotional investment negotiated through the arts can persuade people to take action. Now I want to cite one pertinent example to illustrate this that will be referenced throughout this section and throughout the book.

The London-based creative agency Don't Panic, on behalf of Greenpeace UK, created a YouTube video to mobilise action against the dangerous practices of Arctic drilling. The YouTube short video satirised the title and lyrics of the Academy Award winning Tegan and Sara song 'Everything Is Awesome' from *The LEGO Movie* (2014). Once the short video was uploaded on 8 July 2014, it went viral with over 1 million views in the first day, and has since attracted 8 million views by early 2017. Entitled *LEGO: Everything Is NOT Awesome*, the Greenpeace 1 minute 45 second video displays many LEGO characters and sets living in an Arctic setting.[3]

As the song plays, which is a slower, more methodical version of the original fast-paced song in the film, the camera shots follow an entire LEGO community (in still motion) living and thriving in the Arctic. It shows the community playing football (soccer) and hockey, as well as polar bears and Arctic wolves on ice banks (white LEGO blocks). The short video then shifts to an offshore oil rig next to the community with a Shell flag on top and a corporate oil tycoon smoking a cigar and smirking on the platform (see Figure 2.1). At this moment, the

Figure 2.1 'Oil Tycoon on Platform'. *LEGO: Everything Is NOT Awesome* (Don't Panic and Greenpeace UK, 2014), with kind permissions from Greenpeace.

base of the rig begins to visibly leak crude oil at the ocean surface. Viscous oil slowly pours onto the LEGO set, gradually flooding the LEGO people, animals, buildings, and landscape (see Figure 2.2 and 2.3) as the song continues to remind viewers how 'everything is awesome'. Oil even submerges Santa and an elf. In a matter of 30 seconds, oil has completely replaced the Arctic setting. By the end of the video, the Shell flag at the top of a hill is the only LEGO piece not immersed in the crude oil spill. Appearing next to the Shell flag is the phrase: 'Shell is polluting our kids' imaginations'.

Figure 2.2 'Family in Oil'. *LEGO: Everything Is NOT Awesome* (Don't Panic and Greenpeace UK, 2014), with kind permissions from Greenpeace.

Figure 2.3 'Boy Crying'. *LEGO: Everything Is NOT Awesome* (Don't Panic and Greenpeace UK, 2014), with kind permissions from Greenpeace.

The aim of *LEGO: Everything Is NOT Awesome* was to draw on the artistic and imaginative appeal of LEGO toys and *The LEGO Movie* to confront possible environmental destruction of the Arctic by pressuring LEGO to end its long-time partnership with Anglo-Danish company Shell. In the video description, Greenpeace states:

> We love LEGO. You love LEGO. Everyone loves LEGO. But when LEGO's halo effect is being used to sell propaganda to children, especially by an unethical corporation who are busy destroying the natural world our children will inherit, we have to do something. Children's imaginations are unspoilt wilderness. Help us stop Shell polluting them by telling LEGO to stop selling Shell-branded bricks and kits today. Greenpeace is calling on LEGO to end its partnership with Shell to Save the Arctic.

The problem is not only that LEGO uses petroleum in their plastic toys. Shell is also a sponsor of LEGO, with many Shell logos plastered on the toymaker's kits.

Because of the global popularity of *The LEGO Movie*, a petition circulated calling on the Danish company LEGO to forgo its relationship with Shell, a partnership that dates back to the 1960s, and eventually remove petroleum from their products. The result of *LEGO: Everything Is NOT Awesome*, drawing on people's enormous support of *The LEGO Movie*, was that LEGO dropped its multimillion-pound deal with Shell. Although LEGO issued statements that Greenpeace should 'have a direct conversation with Shell', the toy company eventually capitulated and now promotes its 'green credentials' by 'looking for alternatives to the crude oil from which it currently makes its bricks' (Vaughan 2014). In 2015, LEGO announced on their website they were investing 1 billion Danish krone to search for sustainable materials as a substitute to plastic. A group of over 100 employees were part of the global initiative to establish the LEGO Sustainable Materials Centre (formed in 2015). With over 60 billion LEGO pieces made in 2014 alone, the goal is to replace all oil-based plastic materials by 2030 (Trangbæk 2015). Perhaps not coincidentally, these monumental changes to LEGO's corporate environmental responsibility occurred a year after the Greenpeace UK campaign with the creative argument of *LEGO: Everything Is NOT Awesome*.

As the LEGO example illustrates, creative responses about and from society confront environmental issues by overlapping modes of print, audio, and visual texts. *The LEGO Movie*, *LEGO: Everything Is NOT Awesome* (itself a creative work of media with print and visual texts), and LEGO toys all attempt in various ways to reproduce the narrative about how we use oil in society. As stated at the end of the Greenpeace video, 'Shell is polluting our kids' imagination' (see Figure 2.4). This suggests that petroleum pollutes LEGO toys, which in turn pollutes imaginations of children using them, as much as the environments in which they live. It also implies that the narrative and perception of oil in our everyday lives poison the imagination as much as the environment.

If we unpack the lyrics of the song, even without the visual elements, viewers are expected to tacitly accept the following lyrics: 'Everything is awesome. / Everything is cool when you're part of a team. / Everything is awesome, / when you're living on a dream'. Everything *is* awesome *only* when it is a dream, hence it is not very awesome in reality. Beyond the obvious satire of the lyric contrasting with the annihilation of the LEGO community, the song and film emphasise that being a team improves society. The film short confirms this by including LEGO pieces playing and working together while smiling. Shell, however, is not part of

Figure 2.4 'Shell Is Polluting Our Kids' Imagination'. *LEGO: Everything is NOT Awesome* (Don't Panic and Greenpeace UK, 2014), with kind permissions from Greenpeace.

this team; it is the antagonistic element in the video represented through images of pollution and oppression, which is ironically juxtaposed against the song's lyrics that promote social unity.

The point in explaining the LEGO controversy is to suggest that the way we perceive, narrate, and value both real and imaginative spaces influence the way we identify our physical and creative relationships to environments. This mobilising form of social action can be achieved through print and visual texts in the humanities. The inter-, cross-, multi-, and transdisciplinary thrust of the environmental humanities may allow for broader access to generate change in society because of the way scholars bring diverse interests and methodologies to a particular subject.[4] The environmental humanities provide, according to Deborah Bird Rose and co-authors, a 'thicker' way of seeing humanity, 'one that rejects reductionist accounts of self-contained, rational, decision making subjects'. Instead, people take part in living ecologies – interconnected and intertwined in a globalised world – that contain meaning and value in the network of culture and history (Bird Rose et al. 2012, 2). The following chapters in this book, similar to the analysis of *LEGO: Everything Is NOT Awesome*, provide interdisciplinary examples through specific creative works in literary and visual culture to increase the clarity of focus and spotlight a 'thicker' way of seeing the humanities connect to spatial issues affecting environments. Returning to Morton's quote at the start of this section, in order to have an environment, there needs to be an idea and place for it. *LEGO: Everything Is NOT Awesome* is one instance of where an idea of resistance can be located in a particular place and one that relies on creative engagement to mobilise social and environmental change. This approach exemplifies the major goals of the environmental humanities.

Why do we need the environmental humanities when there already exists an established method of ecocriticism or other social responses of environmental justice? Ecocriticism originated as a way for literary and cultural scholars to investigate the global ecological crisis through the intersection of literature, culture, and the physical environment. As Cheryll Glotflelty memorably writes, 'ecocriticism takes an earth-centred approach to literary studies' (1996, xviii). In the 1970s, scholars wanted to use literary analysis rooted in a culture of ecological thinking, which would also contain moral and social commitments to activism (Meeker 1972; Rueckert 1996). However, its original aim focused on literary, historical, and cultural dimensions of environmentalism in mostly anglophone American and British contexts. Although it remains by definition largely literary and cultural in scope, ecocritical scholarship is no longer limited to these critical terrains and is now more invested in broader approaches in the environmental humanities. Because of its size and influence in the academy, ecocriticism is often used as a catch-all term for parts of the humanities that confront environmental issues or even describe ecologies related to built and non-built environments. Work in environmental justice looks more broadly at how social movements mobilise against ecological damage rather than at the creative works that confront these issues. In themselves, these methodologies provide intra-disciplinary perspectives (i.e. working within a single discipline even if informed by others). The environmental humanities combine these approaches, while also adding the possibility for deeper and 'thicker' scrutiny from an even wider range of perspectives and expertise.

The environmental humanities have arisen out of the reactions of critics, students, artists, writers, and scholars in multiple disciplines and communities to the substantial need to address global ecological transitions and futures. Confluences of environmental thinking and the humanities have existed for a few decades, where indigenous, postcolonial, literary, feminist, and ecological studies have examined the links between oppressive social structures and environmental destruction, but the term 'environmental humanities' originated in 2001 by the Australian historian Libby Robbin, ethnographer Deborah Bird Rose, and ecofeminist Val Plumwood (Adamson 2017, 5). The aim of the environmental humanities is to understand the relationship between human responsibility and the perception and imagination of art as they relate to surrounding environments through sustainable social action. The reason for citing *LEGO: Everything Is NOT Awesome* as an example throughout this section is to concretise how multiple platforms in the humanities – such as a multimedia activist campaign correlated to a Hollywood film and responding to an ecological justice issue through music – might be considered through the multidimensional lens of the environmental humanities.

Despite the evidence of successful action, the strengths of the humanities are less utilised to combat strategic change in public policy or the sciences. Further, the transdisciplinary link between the sciences and humanities remains somewhat vexed. In their multi-authored 'Manifesto for Research and Action', the Humanities for the Environment (HfE) global network acknowledges that while 'science is able to monitor, measure and to some extent predict the bio geophysics of global change', its 'analytical power' falls short of understanding the main vector of planetary change, otherwise known as 'the human factor' (Holm et al. 2015, 979). The stylistic structure of the group's name, HfE, tellingly alludes to this non-anthropocentric focus, with the humanities serving *for* the environment rather than dominating it.

The values, beliefs, and organisational systems of human societies, correlated with their vast abilities to alter the Earth through industry and technology, often function outside of scientific calculation and observation. STEM subjects (science, technology, engineering, and math) can predict changes in climate systems and estimate potential outcomes and solutions based upon current conditions of the environmental carrying capacity of the Earth, which is used as a way of understanding the limits of human impact across time and scale of the Earth. The unpredictability of human endeavour, however, becomes the scientific expanse where research in the humanities excel. While STEM subjects measure and document variables of scale, offering structural solutions, they do not examine social or cultural change to combat the causes of the initial problem (e.g. social and spatial formations). This returns us back to Foster's claim, 'the crisis of the earth is not a crisis of *nature* but a crisis of *society*' (1999, 12; original emphasis). Humans draw on language, narrative, imagination, and cognitive models to conceptualise problems and possibly take action (Holm et al. 2015, 981). Joni Adamson confirms in the introduction to her co-edited anthology *Humanities for the Environment*: 'in the first decade of the twenty-first century, scientists, policy-makers, business and education leaders are declaring the "environmental humanities" crucial to addressing the anthropogenic factors contributing to dramatic environmental changes' (2017, 4). Both the humanities and the STEM subjects are equally essential components to understanding and reversing the environmental impacts rapidly altering the Earth, but the humanities more specifically address the social systems formed through perception and values. This revelation has led some to reconfigure STEM as STEAM (including the arts).

While the 'human factor' is a fundamental aspect to the environmental humanities, it does not function as the dominant subject of scholarship. The environmental humanities, despite the potentially misleading name, attempt to minimise the human subject. Decentring human species while elevating the nonhuman (e.g. animals, trees, and posthuman forms of artificial intelligence) has become one of the key conceptual shifts in environmental theory since the turn of the twenty-first century. Such an inclusive perspective allows space for other narratives to develop alongside what has historically been more anthropocentric. Many researchers in the humanities now recognise the need for interrelated approaches stressing relationships among nonhumans and humans, such as Earth-centric perspectives of biocentrism; emphasising the agency of other species; drawing on material studies of matter and nonhuman agency; investigating artificial life (or AI) and intelligence related to machines and the 'end of nature'; emphasising system/network theory; and understanding asymmetrical relationships between objects in object-oriented ontology (OOO). Work in the environmental humanities moves well beyond antiquated studies of socially constructed concepts of nature from a human perspective, even though humans continue to be central because of their role in ecological destruction. For my own purposes, the 'human factor' is a primary focus because of the ways humans socially produce and construct damaging systems to the biosphere, which produces harmful results for other humans and nonhumans. Humans also produce creative works of art that respond to these affects throughout history.

One way to define and understand the human factor as the major vector of biophysical and geologic change is through the concept of the 'Anthropocene epoch' or the 'age of the human' (*c.* 1784–present). The Anthropocene, which is discussed more at length in the next chapter, underscores 'the human' as the primary motivator of geological and planetary change since the beginning of the age of industry in the late eighteenth century. A second phase of the Anthropocene entails a 'Great Acceleration' of human advancement in technology and consumption over the last 70 years since the 1940s, leading to an alarming rise in greenhouse gases that have triggered destructive weather patterns known as climate change. The humanistic element is an entry point, one of relationality, rather than one of privilege. It underscores the diversity of the humanities, with many interconnected and overlapping methodologies and foci to contend within any situation, but privileging no discipline or humanistic endeavour over another. All of these factors point to humans, but without implying hierarchy, as the origin and potential solution of the current ecological crisis.

In another poem from *The Guardian*'s 'a climate change poem for today' series, Welsh poet Gillian Clarke captures the indifference of the Earth to the human factor in 'Cantre'r Gwaelod' ('The Drowned Hundred'). Focused on a once vaulted place now lost under Cardigan Bay in Wales, the poem describes the petrified forest that becomes unearthed after the storms of February 2014. One concentration of the poem is on the nonhuman material memory of the beach at Borth, which 'is a graveyard, a petrified forest . . . drowned by the sea six thousand years ago'. The beach at Borth with 'stilted walkways' slowly 'tell their story' of 'how they walked over water between trees' when the 'world turns reflective'.

The other aspect of the poem highlights the irrelevance of humans in this deep geologic history. The poem concludes with the following stanza:

> for Earth's intricate engineering, unpicked
> like the flesh, sinews, bones of the mother duck
> crushed on the motorway, her young
> bewildered in a blizzard of feather;
> the balance of things undone by money,
> the indifferent hunger of the sea. (Clarke 2015)

The poem recognises Earth's agency, separate from but supportive of human existence. While humans construct notions of place, they do so to inhabit Earth, despite its indifference towards human survival. The Anthropocene is a 250-year narrative about human's immense influence over the planet. This period holds humans responsible for their own survival, but it also underscores interconnectivity among many other nonhumans. Elsewhere in the poem, it states 'how it will be as world turns reflective' grieving 'for lost wilderness'. Memory dominates the poetic lines, but it is from the perspective of the Earth, one of the petrified forest off of the beach at Borth, not one of human prominence.

In addition to media campaigns such as *LEGO: Everything Is NOT Awesome*, poems such as 'Cantre'r Gwaelod' serve as influential factors in reproducing artistic and social spaces. This poem narrates the environmental experience from the perspective of material landforms on Earth, thereby decentring the human in the formal spaces of the actual poem and in the deep geological history the poem occupies. Clarke's 'Cantre'r Gwaelod' is a poem that eschews human privilege while also engaging with the human factor.

Developing multidisciplinary methodologies of the environmental humanities provide a foundation of understanding and critical expression about ameliorating the drastically 'changing Earth system'. Examining specific ways of responding to what are the 'environmental challenges of our time', according to some of the founders of the environmental humanities movement such as Deborah Bird Rose and Thom van Dooren, among other co-authors, signal how 'the environmental humanities engage with fundamental questions of meaning, value, responsibility and purpose in a time of rapid, and escalating, change' (Bird Rose et al. 2012, 1). Tobias Boes further proposes that we need to develop 'a set of hermeneutics and poetics (a theory of understanding and expression) that might accompany the scientific study of the changing Earth system' (2014, 168). Boes' suggestion would speak to the critical and pedagogical aspects of the environmental humanities that accompany other approaches to theory and activism. Albrecht's use of solastalgia provides a similar basis for theorising understanding, expression, and experience related to environmental damage, even though he frames it as a psychoterratic terminology and practice.

Environmental Humanities is an international and open-access journal now with Duke University Press publishing interdisciplinary scholarship that may not fit within established environmental sub-disciplines. In the 2012 introductory issue, the group of co-authors led by Bird Rose identify three central questions underlining the environmental humanities. First, how do we 'think through' the environment? In what ways can we narrate, perceive, interpret, or acknowledge environmental concerns? Second, how can we unsettle 'dominant narratives' through criticism, theory, pedagogy, as well as other practical forms of action, resistance, and activism? The objective here is to identify the balance between critique and action, which raises the question: how might an academic methodology or creative endeavour such as writing or filmmaking support this balance? Third, how do we create 'bridge-building' among contrasting and contradictory narratives in such instances as community-building or social impact? (Bird Rose et al. 2012; Hutchings 2014.)

There are many evident overlaps between the purposes of spatial justice in geographical theory and practice and those in the environmental humanities. They both 'think through' how we might perceive or interpret environmental and spatial concerns – particularly in how we reclaim and construct narratives about socially produced spaces. Who controls these narratives? Who can circulate and promote them? Looking back to *LEGO: Everything Is NOT Awesome*, the goal is to specifically reclaim the narrative about drilling for oil in the Arctic by constructing new stories or ways of perceiving the situation through a democratised open-access media platform. Is drilling in the Arctic beneficial or destructive? Who controls which narrative? The film short also questions the scale of the problem by deploying small LEGO pieces in an otherwise massive Arctic setting. The film thinks through this issue by portraying a result that alters values through a story about destruction.

Another overlap is how they both challenge dominant narratives of power through theory, pedagogy, criticism, and action – which is to say, there is an interconnected link between thinking, teaching, and activism. Finding an effective balance between critique and action is difficult and continues to be perhaps the most challenging aspect of research in spatial justice and the environmental humanities. Many academics fear that the optics of activism might denigrate their academic credibility, compromising analytical research for potentially incompatible social action (Holm et al. 2015, 985). Equally, many creative artists are careful of the possible reductive ends of their work that activism might publicly invoke. Forms of artistic production that take a position could be perceived negatively as didactic and overstated.

Challenging dominant narratives of power necessitates many approaches to both analytical and conceptual thinking, drawing on both structure and agency inherent in place studies. The Greenpeace video campaign against LEGO and Shell challenges the situation through visual culture and music, while it also persuades the viewer to take action by asking people to sign the petition at the end of the video. The video contests dominant narratives about oil – which attempt to reinforce how oil is economically essential and life-sustaining – by showing how drilling for oil in the Arctic will pollute water sources, melt polar ice, and irrevocably destroy habitats and food supplies for animals and organisms, in addition to releasing more carbon into the atmosphere and increasing global temperatures. The visual and auditory criticism in *LEGO: Everything Is NOT Awesome* segues effortlessly from conceptual thinking to taking action.

Finally, the methodologies employed in the environmental humanities and spatial justice attempt to 'build bridges' among communities, government, and industry as both academic and social modes of engagement. What are the synergies between various levels of policy and institutional learning? How can these links be expanded to influence people's perceptions and values about the ways in which we confront environmental injustices? One of the notable results of *LEGO: Everything Is NOT Awesome* is that it succeeded in its activist aim of building bridges with LEGO and influencing social action and subsequent change. While LEGO openly admitted the battle should be between Greenpeace and Shell, they nevertheless attempted to appease its supporters and consumers by looking for an oil replacement. As previously mentioned, LEGO has been developing a sustainable approach to the plastic bricks that removes any use of oil since 2015 (a year after the successful Greenpeace campaign and viral video in 2014). This caused another possible ripple effect because Shell, among other multinational petroleum companies, wanted to drill in the US Arctic. After this Greenpeace campaign, whether there was a direct connection or not, the Obama Administration placed a temporary moratorium on Arctic drilling in 2016.

Constructing a productive dialogue between humanistic disciplines, industry, and businesses continues to be the most difficult obstacle in 'building bridges'. But as *LEGO: Everything Is NOT Awesome* positively demonstrates, bridges were established and then crossed through the film industry, media, non-profit activism organisations, and businesses. Such an overlap of social process and spatial forms develops from creative and critical work in the environmental humanities, particularly in the ways literature, filmmaking, and media accentuate relationships between social and ecological concerns through narratives of real and imagined spaces.

'Exiles in our own country'

In an attempt to assemble ecological approaches in the humanities to instances of solastalgia caused by spatial injustice, the last part of this chapter investigates Eavan Boland's place-based poem 'In Our Own Country', accompanied by Oliver Comerford's short film *Distance*. Together as overlapping poetic and visual texts, they negotiate the environmentally affected spaces in Ireland through what the geocritic Bertrand Westphal calls stratigraphic vision, which serves as a way of examining place through layers of pasts, histories, archaeologies, and memories. The multimedia format of poetry and film enhance one's attachment to specific places in order to affect social values and perception of place by connecting the viewer through a virtual Web-based experience.

The poems in Boland's collection *Domestic Violence* (2007) are about charged spaces where people live, particularly the domesticated suburban environments developed during the economic 'boom' years in Ireland known as the Celtic Tiger (*c.*1994–2008).[5] Bethany Smith describes Boland's collection as 'archaeological' and probing for 'layers of significance' (2013, 218). The poem 'In Our Own Country' exemplifies the idea of solastalgia because it unravels the archaeological layers of meaning, both in terms of the environmental impacts and the spatial injustices of dislocation and deterritorialisation in Ireland. Boland's poem reveals the tension of past and present in these charged spaces by questioning the development of 'a new Ireland', resulting from neoliberal economic policies supported by the European Union (EU) and multinational businesses relocating in Ireland.[6] Such 'a new Ireland' in the poem, however, is juxtaposed against an older Ireland, which contains similar problems of exile, dislocation, and emigration.

As an Irish artist engaged with the transformations of place, Comerford finds the real and imagined distances between the local and global, predominantly in relationship between the social and environmental. Much of his work explores the concept of place; it attempts to understand place through the visual mediums of film and photography. Filmed in the winter of 2004, *Distance* visualised roads in 19 different counties throughout Ireland. *Distance* pictures the spatial effects that occur in the poem, while it also confronts policies for developing and commercialising a 'new Ireland'. Even the film's name underscores the spatial qualities that separate or dislocate people from place.

Both 'In Our Own Country' and *Distance* lament the loss of place – displayed through various forms of spatial injustice – that ultimately affect the local environments and the people who remain living in them. The multimedia experience overlapping poetry, film, and audio on an open-access website allows for global viewers to also lament the loss of local place through their connective and collective experiences watching the video on the website. This seemingly detached process of Web-based consumption allows a wider audience to experience, even if on a virtual level, instances of solastalgia for the people who have lost a sense of place in Ireland and who remain ecological exiles in their own country. The combined filmpoem format, a medium also discussed in Chapter 4, provides direct exposure for global viewers through online open access – what an isolated poem in an anthology would likely not achieve. The combination of the film and poem creates wider awareness about what might only initially appear as a local issue, but one that resonates universally.

Stratigraphic vision, one of the four fundamental pillars in Westphal's book *Geocriticism*, identifies layered and accumulated pasts, histories, and memories that all construct a given place in space and time. For Westphal, stratigraphic vision relies on temporal factors when reading spaces in literary texts. He argues:

> The diversity of temporalities that we perceive synchronously in several different spaces, even in a single space, is also expressed in a diachrony. Space is located at the intersection of the moment and duration; its apparent surface rests on the strata of compacted time arranged over an extended duration and reactivated at any time. This present time of space includes a past that flows according to a stratigraphic logic. (Westphal 2011, 137)

Here, Westphal suggests that time impacts the perception of space by challenging the moments of both representation (in the text) and real time (in reality). Stratigraphic logic, then, provides a way to examine time (histories or memories) in geographical spaces through literary texts that simultaneously confuse and illuminate spatio-temporal dimensions. If we apply this concept to the filmpoem 'In Our Own Country' / *Distance*, we can see the stratigraphic elements at work through a 'strata of compacted time'.

Both 'In Our Own Country' and *Distance* contain a timeless quality, while they also underscore a clear triad demarcating past and present. The imbrication of past and present in the filmpoem create a simultaneity of what Soja has called the 'both/and also' as a 'Thirdspace' or what Lefebvre describes as a space where the 'past leaves its traces' is 'always, now and formerly, *a present* space, given as an immediate whole' (Soja 1996, 3; Lefebvre 1991, 37; original emphasis). Rather than adopting reductionist epistemologies of 'master narratives' or 'totalising discourses' (producing an either/or choice), often drawn from imperialist ideologies, the 'both/and also' logic associated with Soja's Thirdspace (or whatever term one chooses to use) draws from postmodern theories of knowledge formation that seek to alleviate deep divisions within spatial and historical analysis by creating simultaneous possibilities. Thirdspace is a flexible way of thinking about fluctuating ideas, events, appearances, and representations, and how these affect the material and perceptual ways geographical spaces change (Soja 1996, 3). The time-space dimension is interlinked, presented as both a product and process unfolding out of multiple histories and scales, and even as possible futures. These spatial descriptions by Lefebvre and Soja stress the stratigraphic logic of both the temporal and spatial dimensions in Boland and Comerford's combined filmpoem.

The speaker of the poem establishes the spatio-temporal layers from the beginning: 'An old Europe / has come to us as a stranger in our city, / has forgotten its own music, wars and treaties, / is now a machine from the Netherlands or Belgium'. The speaker describes historical movement between the past – an old 'forgotten' Europe with its legacies of imperialism and colonisation – and the present 'machine' running the new Ireland at the EU headquarters in Brussels. The older legacies of colonialism with its 'music, wars and treaties' transition in this stanza both in time and space to contemporary forms of neocolonialism in a new Ireland made possible by neoliberal globalisation and uneven geographical and economic development. The speaker underscores geographical movement of displacement and domestic forms of exile 'as a stranger in our city'.

Boland also acknowledges how roads perform as transitions and mark points in Irish history witnessed in the present. Roads are vectors of stratigraphic logic that

Figure 2.5 'Road at Night with Arc Lamps'. *Distance* (Oliver Comerford, 2004), with kind permissions from Comerford.

not only link space-time, but also underline the 'perceived-conceived-lived triad' outlined by Lefebvre in the last chapter that signals real and imagined 'lived' space (1991, 40). While roads offer accessibly and mobility across geographies, a metaphor of progress, they also create perceptions of stasis and suspended time. The poem's opening two stanzas mark movement through space and time using the notion of the road as the simultaneous perceived-conceived-lived triad: 'They are making a new Ireland / at the end of our road / under our very eyes, / under the arc lamps they aim and beam' (see Figure 2.5). The poem then breaks into a new stanza, switching into the personal voice using the collective pronoun 'we' of the people affected: 'into distances where we once lived / into vistas we will never recognise'. The focus of this combined text centres on spaces – physical and real spaces of the 'vistas' and 'roads' in Ireland – but also the imagined spaces of new and old Ireland overlapping and yet separate, confused by history and memory. The line 'vistas we will never recognise' signals the collective experience of solastalgia. While the speaker of the poem (speaking on behalf of local people) explains the memory of certain vistas shifting to new panoramas, the speaker also laments this change as a person who must continually witness these preventable changes in a new Ireland.

The interplay between visual media and poetry creates additional effects that extend beyond both mediums individually. For example, *Distance* interprets the poem by envisioning through film what loss of place in Ireland might *look* like 'under our very eyes'. If we move back to the opening lines of the poem, the film captures the comparison between the new and old Ireland through two similarly juxtaposed images. The opening begins with a point of view shot in a car down a rural road. This initial establishing shot only contains the diegetic sounds of the car on the road without any of the speaker's (Regina Crowley) voice-over reading. The film then cuts smoothly to a similar point of view shot of another road, only this road is at night in an unknown city 'under the arc lamps'. Once the contrasted shot of the city appears, the speaker's voice begins to read the opening lines of the poem. The real spaces in *Distance* enhance the imagined qualities in the poem, thereby creating connection with the viewer on multiple scales and through various texts. The film's pathos appears through the emptiness of never-ending footage of roads leading nowhere and parallels the poem's refrain of being 'exiles in our own country', trapped in this never-ending cycle of roads while witnessing the transformed 'vistas'.

In *Distance*, travelling upon roads accentuates the spatio-temporal movement through stratigraphic logic because the viewer experiences images of a road moving across histories and topographies that confuse linear time. There are only seven cuts in the entire three-minute film. Each film cut produces another image of a road from the point of view of a driver/viewer. The roads change between rural and urban, paved and unpaved, night and day. The variations of the roads reveal changes in the scenery, the changed 'vistas', and as a sense of time and movement. Time and space appear interlinked. J'aime Morrison has convincingly argued roads in Ireland function as 'vital cultural spaces' that are 'conduits for movement and memory operating within the larger spatial history of Ireland' (2009, 75). Roads are also markers of direction and certainty symbolising development and progress.

Distance presents roads as another symbol of modernity, supplementing Boland's description of 'making a new Ireland' in the poem, while also underscoring how roads remain integral in Ireland's history. In this sense, roads exemplify stratigraphic vision through memory, history, and territory emerging from Ireland's topographies. While roads allow for greater mobility across topographies, they physically divide and destroy environmental habitats connected to these places. Roads might connect people to spaces of commerce and development, but in so doing they often undermine local relationships between people and environments. A crucial commentary here is that roads regularly function as no-places, serving as a liminal space between places devoid of what Lawrence Buell calls 'associatively think' meanings related to a person's sense of 'place-attachment' or place-home (2005, 63). As the film infers, roads disconnect rather than connect people to place (see Figure 2.6).

Although one of the underpinning themes of the poem reveals charged spaces, the footage in *Distance* sparks an association with the environment by spotlighting recently developed industrial locales in Ireland. An ecological reading of this situation would interrogate how uneven development in globalised capitalism contrasts with policies of sustainable growth. The environmental critique, while firmly articulated in both the poem and film as a loss of place, remains couched in larger stratigraphic visions of space related to urbanisation, dislocation, and exile.

Figure 2.6 'Road as a No-Place'. *Distance* (Oliver Comerford, 2004), with kind permissions from Comerford.

As I have already mentioned, there are two interwoven spatio-temporal views of Ireland in the poem: old and new. The poem's seemingly simple dichotomy underscores two phases of dislocation the speaker discusses in Ireland: old legacies of colonialism and new economic gains that existed during the Celtic Tiger years in the 1990s and 2000s before the collapse in 2008.

Published in 2007, prescient of the global financial crisis of 2008, 'In Our Own Country' confronts the effects of neoliberal globalisation on Ireland's urban and rural geographies. As Jody Allen Randolph points out, the speaker's approach in the poem is not a 'solitary perspective of a nature poet', but the 'communal "we" that is both contemporary and historical' (2013, 169). The aesthetic value of 'nature' redolent in much commentary of Irish poetry shifts in Boland's poem to engage with current political and economic realities facing contemporary society, one of which includes environmental damage. A voice of concern and confrontation supplants praise of Ireland's 'natural' green spaces. The poem achieves this aim by confronting the decimation of place through the space-time layers of stratigraphic vision. It directly addresses the spatial injustices occurring in the new Ireland by capturing a loss of place, one that the speaker bemoans as 'where we once lived', through layers of history and geography.

Ecological themes of uneven development at the centre of globalised economies represented in the poem also correspond with spatial injustices resulting in deterritorialisation, dislocation, and exile. Deterritorialisation strips away any identity attached to a specific place imagined in a territory or nation (Westphal 2011, 144). David Harvey has previously situated this phenomenon of displacement in the context of European colonisation, where planetary spaces 'were deterritorialized, stripped of their preceding signification' only to then be 'reterritorialized' by colonial administrations (1990, 264). Boland parallels dated but still resonant forms of European colonialism with current EU forms of neocolonialism, where policies are constructed to enhance global capital for other EU countries rather than for Ireland. As the poem indicates, the Irish are 'here to watch'; they are 'looking for new knowledge' emerging from under the layers of 'clay'. In order to build one model of a new Ireland, the neocolonial 'machine' must continue unabated, one that results in 'bridge, path, river, all / lost under an onslaught of steel'. The poem implies a contrast where other sustainable and just models of a 'new Ireland' could also be sought. Regardless of economic progress, the speaker in the poem, and the point of view in the accompanying film narrative, lament the present 'new Ireland' – with 'vistas we will never recognise' and the spectral traces of the 'bridge, path, river' that mark the paved over village for development.

What brings together the stratified layers of geography, history, memory, and the ecological diversity now 'lost under an onslaught of steel' remains the spatial attachment to and experience of place. The idea of place-home associated with feelings of solastalgia more accurately reflects what the speaker mourns in the poem. Place carries meaning because it reproduces an emotional, historical, and cultural attachment for people to an existing space, a process not unlike stratigraphic visioning in geocriticism. Place here is layered with meaning and constructed out of social and individual processes.

The poem's closing lines recognise the effects of solastalgia by highlighting dislocation and deterritorialization of people who continually suffer from uneven development in Ireland and also globally. Such an effect creates a new form of symbolic exile where people live in the same geography as before but lose any sense of place-home, thereby becoming 'exiles in our own country'. Through uneven development environments have been stripped of meaning, affecting not only biodiversity of a region, but also the stratified layers of history and memory of an old Ireland into a new Ireland. As the final line of the poem suggests, to be exiled in one's own country equates to being stripped of one's place-home where one feels, returning to Albrecht's quote earlier in the chapter, 'place-based distress in the face of the profound environmental change' (2006, 35). The fused filmpoem 'In Our Own Country' / *Distance* locates a lost place, one that once existed but is now paved over like the roads in a new Ireland. For the speaker of the poem, what remains in this new Ireland is 'nothing, nothing, nothing'.

Notes

1 This differs from being 'out of place', such as literal homelessness or displacement (i.e. refugees), because people suffering from solastalgia remain in lived places regardless of how such spaces have been affected by environmental damage. Previous research has examined transgression of 'moral geographies' that result in people feeling 'out of place',

whether they are refugees who are physically displaced or LGBTQ communities who experience other forms of being 'out of place' in society (Cresswell 1996).

2 There are some notable examples where issues of spatial justice (but without using this terminology) appear in ecocritical (literary) scholarship (Adamson 2001; Buell 2005; Heise 2008; Nixon 2011).

3 I want to acknowledge and thank Danine Farquharson for initially introducing me to this short film, and to both Farquharson and Graeme Macdonald for briefly discussing this film in a workshop we all participated in entitled 'Everything Is NOT Awesome: Thinking through the Energy and Environmental Humanities'. It was co-sponsored by the Edinburgh Centre for Carbon Innovation and the Edinburgh Environmental Humanities Network, and held at the University of Edinburgh on 18 September 2015.

4 For definitions of these approaches, see note 1 in the introduction.

5 The Celtic Tiger was a phenomenon in the 1990s and 2000s ushering in a short era of economic wealth and cultural capital, but also where neoliberal economics exploited previously colonised countries to avail of lower tax rates and a cheaper, educated workforce. The 'boom' (and later bust) resulted in a monumental transformation from a vastly economically depressed Ireland in the 1980s into a decade of inequitable wealth and subsequent uneven development in the 1990s and early 2000s.

6 In reference to the poem 'In Our Own Country', I will be referring to it on the Poetry Project website, which has no page number and is an open-access website, instead of in the poem's original publication (Boland 2007).

Bibliography

Adamson, Joni. 2001. *American Indian Literature, Environmental Justice, and Ecocriticism: The Middle Place*. Tucson, AZ: University of Arizona Press.

——. 2017. 'Introduction: Integrating Knowledge, Forging New Constellations of Practice in the Environmental Humanities'. In *Humanities for the Environment: Integrating Knowledge, Forging New Constellations of Practice*, edited by Joni Adamson and Mike Davis, 3–19. New York: Routledge.

Albrecht, Glenn. 2005. 'Solastalgia: A New Concept in Health and Identity'. *PAN: Philosophy, Activism, Nature* 3: 41–55.

——. 2006. 'Solastalgia: Environmental Damage Has Made It Possible to Be Homesick without Leaving Home'. *Alternatives Journal* 32(4/5): 34–6.

——, Gina-Maree Sartore, Linda Connor, Nick Higginbotham, Sonia Freeman, Brian Kelly, et al. 2007. 'Solastalgia: The Distress Caused by Environmental Change'. *Australasian Psychiatry* 15(1): 95–8.

Allen Randolph, Jody. 2013. *Eavan Boland*. Lewisburg, PA: Bucknell University Press.

Bird Rose, Deborah, Thom van Dooren, Mathew Chrulew, Stuart Cooke, Matthew Kearns, and Emily O'Gorman. 2012. 'Thinking Through the Environmental, Unsettling the Humanities'. *Environmental Humanities* 1: 1–5.

Boes, Tobias. 2014. 'Beyond Whole Earth: Planetary Meditation and the Anthropocene'. *Environmental Humanities* 5: 155–70.

Boland, Eavan. 2007. *Domestic Violence*. Manchester: Carcanet Press.

Buell, Lawrence. 2005. *The Future of Environmental Criticism: Environmental Crisis and Literary Imagination*. Cambridge, MA: Blackwell.

Casey, Edward. 1993. *Getting Back into Place: Toward a Renewed Understanding of the Place-World*. Bloomington, IN: Indiana University Press.

Comerford, Oliver. 2004. *Distance*. Dublin: Department of Communications, Marine and Natural Resources.

Cresswell, Tim. 1996. *In Place/Out of Place: Geography, Ideology, and Transgression*. Minneapolis, MN: University of Minnesota Press.

Clarke, Gillian. 2015. 'A Climate Change Poem for Today: Cantre'r Gwaelod'. *The Guardian*, 19 May 2015. Accessed 31 January 2017. www.theguardian.com/environment/2015/may/19/a-climate-poem-for-today-cantrer-gwaelod-by-gillian-clarke.

Dorgan, Theo. 2015. 'A Climate Change Poem for Today: The Question'. *The Guardian*, 1 June. Accessed 12 December 2015. www.theguardian.com/environment/2015/jun/01/a-climate-change-poem-for-today-the-question-by-theo-dorgan.

'Drone researchers show extent of "Desmond" floods'. 2016. University of Salford, Manchester. Accessed 20 January 2017. www.salford.ac.uk/news/articles/2016/drone-researchers-show-extent-of-desmond-floods.

Foster, John Bellamy. 1999. *The Vulnerable Planet: A Short Economic History of the Environment*. New York: Monthly Review Press.

Glotfelty, Cheryll. 1996. 'Introduction: Literary Studies in an Age of Environmental Crisis'. In *The Ecocriticism Reader: Landmarks in Literary Ecology*, edited by Cheryll Glotfelty and Harold Fromm, xv–xxxvii. Athens, GA and London: University of Georgia Press.

Harvey, David. 1990. *The Condition of Postmodernity: An Enquiry into the Origins of Cultural Change*. Oxford: Blackwell.

——. 1996. *Justice, Nature, and the Geography of Difference*. Oxford: Blackwell.

Heise, Ursula. 2008. *Sense of Place and Sense of Planet: The Environmental Imagination of the Global*. Oxford: Oxford University Press.

Holm, Poul, Joni Adamson, Hsinya Huang, Lars Kirdan, Sally Kitch, Iain McCalman, et al. 2015. 'Humanities for the Environment: A Manifesto for Research and Action'. *Humanities* 4: 977–92.

Hutchings, Rich. 2014. 'Understanding of and Vision for the Environmental Humanities'. *Environmental Humanities* 4: 213–20.

Lefebvre, Henri. 1991 [1974]. *The Production of Space*. Translated by Donald Nicholson-Smith. Oxford: Blackwell.

LEGO: Everything Is NOT Awesome. 2014. Directed by Andy Gent. London: Don't Panic, Unit9, and Greenpeace. YouTube video, 1:44. Posted 8 July. www.youtube.com/watch?v=qhbliUq0_r4.

Meeker, Joseph. 1972. *The Comedy of Survival: Studies in Literary Ecology*. New York: Scribner's.

Morrison, J'aime. 2009. '"Tapping Secrecies of Stone": Irish Roads as Performances of Movement, Measurement, and Memory'. In *Crossroads: Performance Studies and Irish Culture*, edited by Sara Brady and Fintan Walsh, 73–85. Basingstoke, UK: Palgrave Macmillan.

Morton, Timothy. 2007. *Ecology without Nature: Rethinking Environmental Aesthetics*. Cambridge, MA: Harvard University Press.

Nixon, Rob. 2011. *Slow Violence and the Environmentalism of the Poor*. Cambridge, MA: Harvard University Press.

Pidd, Helen. 2015. 'Storm Desmond: Lancaster's Small Businesses Face "Zombie Apocalypse"'. *The Guardian*, 6 December. Accessed 9 January 2017. www.theguardian.com/world/2015/dec/06/storm-desmond-lancasters-small-businesses-face-zombie-apocalypse.

Rueckert, William. 1996 [1978]. 'Literature and Ecology: An Experiment in Ecocriticism'. In *The Ecocriticism Reader: Landmarks in Literary Ecology*, edited by Cheryll Glotfelty and Harold Fromm, 105–23. Athens, GA and London: University of Georgia Press.

Soja, Edward. 1996. *Thirdspace: Journeys to Los Angeles and Other Real-and-Imagined Places*. Malden, MA: Blackwell.

——. 2009. '"The City and Spatial Justice / La ville et la justice spatiale". Interview with Sophie Didier and Frédéric Dufaux'. *Spatial Justice / Justice Spatiale* 1: 1–5.

Smith, Bethany J. 2013. 'Ekphrasis and the Ethics of Exchange in Eavan Boland's *Domestic Violence*'. *Word & Image: A Journal of Verbal/Visual Enquiry* 29(2): 212–32.

Trangbæk, Roar Rude. 2015. 'LEGO Group to Invest 1 Billion DKK Boosting Search for Sustainable Materials'. LEGO website, 16 June. Accessed 2 January 2017. wwwsecure.lego.com/en-us/aboutus/news-room/2015/june/sustainable-materials-centre.

Vaughan, Adam. 2014. 'LEGO Ends Shell Partnership Following Greenpeace Campaign'. *The Guardian*, 9 October. Accessed 2 January 2017. www.theguardian.com/environment/2014/oct/09/LEGO-ends-shell-partnership-following-greenpeace-campaign.

Westphal, Bertrand. 2011. *Geocriticism: Real and Fictional Spaces*. Translated by Robert T. Tally Jr. London: Palgrave Macmillan.

Whitehead, Mark. 2014. *Environmental Transformations*. London: Routledge.

Part II

Oil

3 Petrospaces

Society and oil

Oil is complex. It is not only geophysical and economic, but also social and cultural in the many ways it conceptually and practically influences our lives. And yet, how societies capture and convert carbon-based energy such as oil and gas continue to undermine environmental and social sustainability. The historic relationship between ecological balance and carbon-reliant infrastructures is both a fraught and symbiotic one. As a society, we need forms of energy to support our current way of living, and yet our most consumed form of fossil fuel energy is both toxic and finite. A 2013 report by the World Energy Council projects by 2050 we will need 60 per cent more energy than at our current rate of usage – if, that is, we do not change how and why we use energy (Whitney 2013, 16).

At the same time, the Earth's environments and inhabitants are suffering from increased greenhouse gases, such as carbon dioxide and methane in the atmosphere, resulting in hotter temperatures and subsequent water shortages, with the need for larger, far-reaching energy systems increasing. Levels of carbon dioxide (CO_2) in the atmosphere have increased 40 per cent since the start of the industrial age in the late eighteenth century, but half of this upsurge has remarkably been since 1970 (Mitchell 2011, 7). The rise of CO_2 is the leading cause of the current climate breakdown. If we account for the future proliferation of carbon anticipated by the World Energy Council in relation to our current forms of energy production, we will, as most climate scientists point out, alter the Earth's ability to sustain life.

The story of oil and its confirmed cause of climate change is *the* story of our lifetimes, and part of the story of *Ecological Exile*. Questions about oil and culture in the arts and humanities remain tangential in overall discussions about energy futures, despite the ubiquity of oil culture in our social lives. How do we define and think about oil in culture? How might we understand oil as spatial as well as social? Can or do we develop discourses to understand the pervasiveness of oil culture? As I clarify more extensively in Part III, 'Climate', climate change (global warming) is an issue of spatial justice. This chapter, however, explores spatial approaches to oil, particularly how space reveals ecological and social instabilities related to oil and gas production as they are represented in literary and visual culture. Oil is used in this book (and elsewhere by others) as a metonym for the larger fossil fuel industry (oil, gas, coal, etc.) and it is linked to social formations of the economy, culture, and geography (Szeman et al. 2016, 19). Oil also comprises and shapes space in the cultural imagination.

This opening chapter of Part II, 'Oil', discusses the energy humanities, time-space compression of neoliberalism, and the Anthropocene within the period of modernity referred to as petromodernity. Approaching various oil-related injustices from what I am calling a 'petrospatial' perspective, this chapter surveys the links among petrocultures, space, and injustice as a way to view how the production of fossil fuels are represented in literary and visual texts and in culture more generally. Aligning with the theme of the book, this chapter looks at oil as a spatial and cultural concern framed within the environmental humanities. It begins by explaining the connections between the energy and environmental humanities within the Anthropocene, ending with Jackie Kay's poem 'Extinction' from *The Guardian*'s 'a climate change poem for today' series as an example. Building upon this the first section, I next consider how the effects of neoliberalism (or late capitalism) have created spaces of placelessness and place-taking that contribute to spatial injustices and feelings of solastalgia.

Overviewing this critical direction will inform the following three chapters in Part II, where I will examine narratives about oil and speed in drama in Chapter 4, literary-infused media about oil and gas refineries in Chapter 5, and documentary film within micro-geographies dealing with pipeline injustices in Chapter 6. Oil may have been first discovered in Russia (what is now Azerbaijan) in 1848 and then later in Pennsylvania in the United States in 1859 (Urry 2013, 39), but my focus in Part II will be on its effect upon drama, poetry, media, and documentary film in Scotland and Ireland, two countries with energy histories as part of British industrialisation. This second part of the book ultimately explores the socio-spatial and cultural associations of oil on individuals and communities who experience ecological exile, as they are represented in specific modes of artistic and creative production.

Energy humanities and the Anthropocene

In addition to infrastructure and public policy, social and cultural approaches in the energy humanities offer dynamic ways to explore links between forms of energy and ecological histories and futures. Many works of art, film, and literature within the humanities have the power to transform social values, particularly in the ways we think, feel, perceive, and imagine the spaces in which we live. Our conceptual systems, drawing on metaphor and imagery, define our everyday lives and subsequently the choices we make (Lakoff and Johnson 1980, 3). Dominic Boyer and Imre Szeman, who both introduced the phrase 'energy humanities', intending to bridge the gap between energy and culture by critiquing structures of energy, explain that approaches to energy must challenge the boundaries around disciplines and between academic and applied research. They argue, 'Energy humanists contend that our energy and environmental dilemmas are fundamentally problems of ethics, habits, values, institutions, beliefs and power – all areas of expertise of the humanities and humanistic social sciences' (Boyer and Szeman 2014, 40). The energy humanities explore the link between conceptual and structural systems of energy, thereby advancing our ability to reframe energy as a social and cultural question and take action toward creating generative beginnings by transitioning away from fossil fuels (Szeman et al. 2016, 14).

The energy humanities slightly differ from the conceptual terrain known as 'petrocultures' because the former is less restricted to issues related to fossil fuels of coal, oil, and gas. Nevertheless, both approaches intersect and so it is worth drawing out a few distinctions and overlaps between the two. Petrocultures began as a research group at the University of Alberta in 2011, founded by Imre Szeman and Sheena Wilson, to support 'research on social and cultural implications of oil and energy on individuals, communities, and societies about the world'.[1] There have been three subsequent biennial and internationally focused conferences held in Canada under the name 'Petrocultures' at University of Alberta (2012), McGill University (2014), and Memorial University Newfoundland (2016). The 2018 conference will be held in Glasgow, Scotland.

Petrocultures function similarly to what Ross Barrett and Daniel Worden have called 'oil studies' in their edited volume *Oil Culture* (2014). Akin to petrocultures, oil studies critically engage with 'oil's symbolic life' to explore 'historical, theoretical, and thematic' scholarship about the culture surrounding oil (Barrett and Worden 2014, xxiii). Both petrocultures and oil studies, as well as the more expansive environmental and energy humanities, partly emerged from Amitav Ghosh's admission in his essay 'Petrofiction', where he called out the surprising absence of fiction (and therefore narratives) about fossil fuels in twentieth-century literature and culture (1992, 30). The notion of petrocultures aims to enlarge the relationship between fossil fuels and literary representation (as first proposed by Ghosh) to one that encompasses many forms of cultural and artistic production.

Functioning as a global research cluster and critical mode of scholarship, the study of petrocultures or 'petrocriticism' critiques the relationship between fossil fuels in various phases of capitalism, which are interdependent and inseparable. Petroculture, as Imre Szeman and Jeff Diamanti contend, is a 'global culture we find ourselves in today', where society has been and continues to be 'organized around the energies and products of fossil fuels, the capacities it engenders and enables, and the situations and contexts it creates' (2017). Graeme Macdonald explains petroculture as a 'mode of interpretive critique' within a field of 'cultural practice' that builds social links both inside and outside of the academy (2017, 36). Approaches to petrocultures exist within the broader environmental (and energy) humanities, offering a specific method of 'interpretive critique' particularly on oil culture, as much as they function outside of the environmental humanities, drawing on Marxist, cultural, feminist, or poststructuralist approaches.

Scholars working in petrocultures ask if fossil fuel culture viewed as a 'petrocapitalism' or 'oil capitalism' could be reshaped to support sustainable development, largely through alternative or sustainable forms of energy? The answer to this question often leads to a second question: can we move past an economic and social system dependent upon fossil fuels? The transition between petrocapitalism and another known or unknown economic systems might seem unfathomable, and it is the essence of this enormous question that petrocultures research tasks itself with. There are 'deep ties' between oil capitalism and representations of it in cultural and artistic production (Barrett and Worden 2014, xxiv). Similar to spatial and place justice (as well as histories of the Anthropocene), the real and imagined elements of oil and gas in culture are socially produced, creating experienced, perceived, and lived experiences in social spaces (going back to Lefebvre's notions of space). These effects stress that fossil fuels be examined in society and culture as much as in industry and policy through a petrocultural existence in which we currently live.

One reason to create the energy humanities was to provide a role for humanistic disciplines in what are now STEM (science, technology, engineering, and math) dominated discussions about energy in the public and private sectors. Szeman and Boyer approach energy largely through a critique of the structures of energy, in addition to how conceptual systems might confront such structures. For them, as well as others such as myself, forms of energy might be explored in society and culture similarly to fossil fuels – how they are socially produced and linked to social formations of extractive production of late capitalist forms of neoliberalism. But, the energy humanities provide a broader scope than petrocultures. As both Szeman and Boyer acknowledge in the introduction to their edited volume *Energy Humanities* (2017), there are many ways to talk about energy in the humanities, particularly the abuses of energy. They argue:

> use and abuse of energy have had a significant impact – perhaps *the* most significant impact – on the shape in which we find the planet today. This is especially the case when it comes to the use of fossil fuels – first coal, and then oil and natural gas. (Szeman and Boyer 2017, 1; original emphasis)

People working on the energy humanities examine how the arts and humanities might explain or even reinterpret energy systems. They explore how we might think about, react to, or theorise energy through culture and society as a belief, value, ontology, or knowledge. Termed 'energy epistemologies', Szeman and Boyer emphasise that 'fossil fuels in particular have been surprisingly hard to figure – narratively, visually, conceptually – as a central element of the modern' (2017, 6). This, of course, takes us back to petrocultures as a form of critique aimed at fossil fuels. Extending STEM to STEAM (adding the arts) serves both as an energy metaphor (i.e. the steam engine is chiefly responsible for carbon-based industrialisation) and as a practical and recommended way forward in discussions about energy transitions and futures. Despite the prevalence and control of oil-reliant social and economic systems for over 200 years, fossil fuel forms of energy remain difficult to understand, discuss, and represent in society. Regardless, print and visual texts provide vectors to teach energy epistemologies or 'how to think about energy' across a range of culture in the humanities.

In short, the energy humanities look at all energy systems, not only the cultural and social links to fossil fuel histories (i.e. petrocultures). Alternative forms of energy with low or no carbon output associated with the humanities are beginning to emerge on the critical landscape, gesturing toward solutions to current and future obstacles while also tied to aesthetic practices. The energy humanities offer a framework to discuss alternative energy models, aesthetic ways of presenting energy forms in public, and predicting energy futures.

Looking at the aesthetics of clean energy generators might be one example of innovative research in the environmental humanities. The Land Art Generator Initiative (LAGI) serves as a model for future development because its remit is to provide 'a platform for artists, architects, landscape architects, and other creatives working with engineers and scientists to bring forward human-centred solutions for sustainable energy infrastructures that enhance the city as works of public art while cleanly powering thousands of homes'.[2] In 2014, the LAGI and Refshaleøen Holding in partnership with IT University of Copenhagen, Danish Architecture Centre, and Information Studies at Aarhus University proposed that public art and clean energy intersect to create 'the new smart city'.[3] Under the funding scheme of LAGI, solar energy generators exist in plain sight, in bays or city parts, and function as art as well as fuel producers (see Figure 3.1).

Another model for aesthetic energy futures is the 'Alien Energy: Social Studies of an Emerging Industry' research project, which conducts comparatives analyses of the impacts of renewable energy initiatives in several locations throughout Europe (Denmark, Iceland, and Orkney Islands, Scotland). Spearheaded by the scholar and artist Laura Watts, 'Alien Energy' looks at how forms of green energy interact with humans, technologies, and the environment.[4] These two

Figure 3.1 *Beyond the Wave* (2014). A submission to Land Art Generator Initiative 2014 Copenhagen by Jaesik Lim, Ahyoung Lee, Sunpil Choi, Dohyoung Kim, Hoeyoung Jung, Jaeyeol Kim, Hansaem Kim, with kind permission from LAGI.

intertwined artistic energy futures exist, but social infrastructures supported by fossil fuels continue to dominate, and thus the reason they receive critique in this part of the book.

All of these approaches explore our energy futures in the hope to create clean energy that is more integrated in and controlled by the people rather than in socially produced systems supported by private corporate interests. If '[e]nergy is a vexingly abstract concept', as stated by J.R. McNeill and Peter Engelke in *The Great Acceleration* (2014), then approaches to energy in the humanities would allow us to understand the various sources of energy essential to existence (7). As the above examples demonstrate, many initiatives continue to emerge that address our energy futures in socially and environmentally sustainable ways. Unfortunately, with the fossil fuels of coal, oil, and natural gas comprising 87 per cent of the global commercial energy mix, changing energy systems and social support of them remain one of the largest issues facing humans and nonhumans in what is now known as the Anthropocene epoch (McNeill and Engelke 2014, 10).

Both the energy and environmental humanities come together in the historical arc of the modern period labelled the Anthropocene. Since the nineteenth century, geologists, Earth scientists, and evolutionary biologists have divided the Earth's geologic history into taxonomies (eras, periods, and epochs) based upon the fossil record (McNeill and Engelke 2014, 1). This geologic time scale (GTS) establishes patterns of existence and extinction through chronological and radiometric dating. The suffix *-cene* derives from the word 'new' in Greek, whereas *anthropos* translates as 'human'. The Anthropocene is the story about how the 'new human' created the modern environmental crisis, largely through fossil fuel energy and industrialisation. Despite the root of the name, it is not anthropocentric. Rather, it challenges the centricity of humans in the order of the Earth by making them largely responsible for planetary destruction.

In response to a significant change in the Earth's systems since the mid-eighteenth century, the atmospheric scientist Paul Crutzen (1995 Nobel Prize winner), along with biologist Eugene Stoermer, concluded that the changing composition of the atmosphere, with an alarming rise in CO_2, might alter the Earth's geologic history enough to justify a new epoch. As a result, they defined the Anthropocene as a geological and biophysical epoch influenced and altered by human/anthropogenic activity during modern history. This finite period influenced by *Homo sapiens* involves high CO_2 levels, species extinction, ocean acidification, and weather system disturbances. The Holocene serves as the previous geologic epoch dating back 11,700 years, when the glaciers receded at the end of the Ice Age, allowing human and nonhuman species to proliferate. In contrast, the Anthropocene epoch (according to Crutzen and Stoermer) loosely begins around 1784 in Britain with James Watt's invention and patent of the coal-powered steam engine, a device capable of turning thermal into mechanical power and ushering in a new age of industrialisation. One major by-product of the steam engine was that it emitted large amounts of CO_2 through accelerated industrial production to unsustainable levels in the nineteenth century.

The spirit of the Anthropocene burns with two major accelerants: rising energy demands and exponential use of fossil fuels to meet them, both in the nineteenth (coal) and twentieth centuries (oil and gas). By the 1890s, for example, only half of the world's energy relied on fossil fuels, but by 2015 the demand rose to 80 per cent (McNeill and Engelke 2014, 2). The Anthropocene must also be considered in the arc of modernity. The project of modernity over the past two centuries – known as one of human enterprise, freedom, and technological progress – has been produced by social formations supporting political economies based upon fossil fuels. As Szeman and Boyer recognise, there is a 'strong equation of energy and modernity' that necessitates fundamental understanding of what forces 'have given shape to modernity', a historical period and concept giving birth to rights, freedoms, and innovation, while at the same time allowing an enormous capacity for 'ballooning capitalist economies' that paradoxically challenge progressive social evolution (2017, 1). The climate historian Dipesh Chakrabarty relatedly links freedom born in modernity to energy when he writes the 'mansion of modern freedoms stands on an ever-expanding base of fossil fuel use. Most of our freedoms are energy intensive' (2009, 208). Modernity is a term that suggests an interlocking relationship between the social and historical, space and time, which assumes that humans, cultures, and institutions produce certain conditions (Berman 1982).

Petromodernity is a term used to critique and also explain these contrasting and overlapping layers of modernity through the systems of fossil energy that primarily fuelled it as an idea and reality. The environmental literary historian Stephanie LeMenager broadly defines 'petromodernity' as 'a modern life based in the cheap energy systems long made possible by petroleum'. LeMenager introduces petromodernity along with the newness of 'post-oil criticism', which draws 'upon the work of environmentally sensitive social scientists' to address the 'social affects' of a 'post-petrol future' (2012, 60; see also 2014). We can also factor petromodernity into explanations of modernity mentioned by Szeman, Boyer, and Chakrabarty above, among others – namely, petromodernity stresses the fundamental links between modern life (with all of its paradoxical freedoms and injustices) and fossil fuel energy systems, while it also gestures toward thinking through energy futures.

Less clear, however, are the ways oil and gas influence and infiltrate critical and creative practices in the humanities. If fossil fuels propelled modernity, how do they also fuel culture within this period? How bound up are narratives about oil and gas in our daily lives? If they are so prevalent, then why do they remain elusive in the social discourse? Invoking Frederick Buell's characterisation of 'fossil-fuel culture' as an 'age of exuberance' in 'social life', but one that is constantly 'haunted' by the spectre of 'catastrophe', petromodernity is best considered as a period haunted by anxiety and catastrophe while also producing opportunity and possibility (2012, 276). Such a history has allowed exponential growth of population and technologies, providing enormous opportunity and potential, but at the cost of environmental balance leading to planetary and species devastation. Petromodernity emphasises the extensive link between fossil fuels and modern history because it combines the three overriding factors causing the global ecological crisis: capitalism, modernity, and fossil fuel energy systems. The Anthropocene offers a way to historically and spatially frame this relationship. It considers geologic and environmental factors within the social formations of modernity, and for these reasons the term has stuck – but not without some continued debate.

Some have contested using the monolithic term Anthropocene. The debate is largely about its genesis. Some purport it begins at the period of global European imperialism in the early fifteenth century, at which time the seeds of capitalism were dispersed around the globe. The historian Jason Moore, for example, defines this arc the Capitalocene rather than the Anthropocene because of how capitalism creates cheap 'natures' related to food, energy, and resources, all of which significantly developed centuries before the eighteenth century (2015; 2016). Similar to Moore, the human ecologist Andreas Malm locates the main cause of ecological collapse to an unequitable system of political economy and therefore believes the Anthropocene does not accurately reflect the root of the problem. He argues the climate crisis is not so much 'the geology' of 'mankind, but of capital accumulation', hence the use of Capitalocene (2016, 391; see also Davis 2016). Industrialisation has altered the biophysical aspects of the Earth enough that it remains convincingly the most accurate way of periodising the environmental crisis. In many ways, the Anthropocene and Capitalocene overlap and address similar issues from different vantage points and are therefore more complementary than in disagreement. I elected to use the term Anthropocene in this book mainly because the scientific community of geologists also use it and will likely soon confirm it as a new epoch in the Earth's GTS chart. Despite controversy, the Anthropocene appears as the most universal way to explain humans' largest impact on the Earth in recorded history.

The second and more influential phase of the Anthropocene that we continually live in is called the 'Great Acceleration' (Crutzen and Stoermer 2000, 17–18; Crutzen 2002, 23). It begins at the end of World War II, when scientists in 1945 began measuring the changes in the atmosphere. Two events catalysed these findings: atom bomb testing in Trinity, New Mexico, and the global atmospheric effects of the subsequent two bombs dropped on Hiroshima and Nagasaki later in 1945 (Morton 2013, 7; McNeill and Engelke 2014, 4). Besides massive carbon dispersal within the Great Acceleration, there has been a global increase of radioactive isotopes because of the amount of thermonuclear weapon testing, leaving anyone born after 1963 with radioactive matter in their teeth (Farrier 2016).

Growth during the Great Acceleration remains unsustainable. From 1945 to 2015, over three human generations, 75 per cent of the world's CO_2 loading in the atmosphere has occurred. Motor vehicles have increased in number from 40 million to 850 million, the number or petroleum-based plastics enlarged from 1 million to 300 million tons, and nitrogen in fertilisers (which acidify waterways and oceans) expanded from 4 million to 85 million tons. Such increases must be fuelled by some form of energy, which is why annual global energy consumption has drastically escalated. In 1965, we used 3,813 million tons of oil, whereas by 2013 the quantity swelled to 12,730 million tons. In addition, the human population has tripled from 2 billion in 1930 to 7 billion in 2011 (McNeill and Engelke 2014, 2, 4, 11).

As these figures show, the speed at which growth has occurred is unprecedented in the history of the Earth, deserving of its own epoch focused on the consequential changes generated by anthropogenic forces. The technologies and advances that propelled the Anthropocene epoch might also cause human extinction, pushing the Earth into another geomorphological period. In the words of the environmental writer Gregory Bateson, 'the organism that destroys its environment destroys itself' (1972, 457).

The Anthropocene is ultimately defied by its human dimension, what environmental humanities scholar David Farrier aptly calls a 'sublime force', where humans, rather fearfully and magnificently, have altered geologic cycles through the premature compression of time and space. He cites, for example, that a single mine in Canada's tar sands can move up to 30 billion tons of sediment each year, which doubles the yearly amount of sediment moved by all the world's rivers combined. In another example, he outlines that the sheer weight of fresh water moved across global geographies has altered the rotation of the earth (Farrier 2016). These events would normally take millions of years in deep geologic time. Unlike the Holocene, the Anthropocene represents the decline and possible mass extinction of human and nonhuman species. Buell succinctly calls the Great Acceleration a period of 'exuberance and catastrophe' (2012, 73), which captures the age of extremes we are currently living in, while Timothy Morton goes so far to suggest that the Anthropocene signals the 'end of the world' as a human concept and ushers in a new way to think about the Earth as a non-anthropocentric entity (2013, 7).

Jackie Kay's 'Extinction', one of the poems from *The Guardian*'s 'Climate Change Poem for Today' series, addresses 'the end of the world' within the Anthropocene. The Scottish poet Kay lists in 'Extinction' all the perceived elements that keep us 'safe' because the world is ostensibly 'a dangerous place', but it is framed in the negative – from the voice of an isolationist speaker resistant to change. Relying on satire and irony, Kay writes a list of what to avoid in order to evade extinction. Framed as an anti-progressive complaint, it satirises the view of the 'affected' – that is, those incredulous of ecological or social injustice. It opens by stating: 'We closed the borders, folks, we nailed it. / No trees, no plants, no immigrants'. It then goes through a long list of saying 'no' to anything related to 'loony lefties' ('no' appears 46 times). Kay organises the poem line by line with lists of what not to accept, such as fresh air, birds, bees, polar bears, ice, vegetarians, carbon curbed emissions, CO_2 questions, greens, Brussels, politically correct classes, readers, and 'no sniveling-recycling-global-warming nutters'.

The real extinction exhibited in the poem is tolerance for new ideas and openness to change our future. It holds humans responsible for how some respond to environmental threats by politicising the issue as a binary choice of either/or. Kay's poem of the Anthropocene emphasises humanistic pettiness by framing the speaker's tone in the negative rather than in the affirmative. It concludes by stating in the speaker's voice: 'Now, pour me a pint, dear. Get out of my fracking face'. The obvious double meaning of 'fracking' encapsulates the theme and tone of the poem because it is used both as an invasive and patriarchal reference ('dear') to ongoing debates about hydraulic mining (or 'fracking'), which is currently banned in Scotland. The invasiveness of fracking parallels the assertiveness of the speaker demanding a drink from the 'dear'. The process of fracking involves injecting high pressure streams of liquid into rocks beneath the ground to break open smaller fissures to extract otherwise impossible to access oil and gas. Fracking serves as one of the more polluting and dangerous forms of extraction because the process contaminates ground and surface water used for drinking and agriculture, creates air and noise pollution, and triggers seismic activity by activating dormant faults or even creating new ones. The term 'fracking' in the poem plays linguistically on the term 'fecking', which is a colloquial expletive derived from 'fucking' and used to express contempt. It ultimately ironizes the ecological injustice of fracking.

Even the final expression of the poem drawing on a controversial term 'fracking' around the planet, not only in the UK and Ireland, conjures many injustices connected to oil and gas production in the social sphere. On the one hand, much of the Earth faces species extinction; on the other hand, we have people who do not want to be bothered by what the curmudgeonly speaker of 'Extinction' calls a 'Little man' and 'Little woman' attempting to create change. The ultimate irony is that the speaker of the poem represents the contemporary cause of potential extinction, particularly through inaction and intolerance. The poem mocks how the speaker reaffirms capitalist forms of constructing social spaces in the Anthropocene as empty backdrops for exploration and exploitation.

The Anthropocene epoch, and more specifically the Great Acceleration, emphasises the impact of the 'human factor' on environmental change, which includes social, economic, and cultural developments as outlined in Kay's poem. The poem's closing comments from the speaker indicate a parallel to the Andropocene, where systemic northern white masculine entitlement has devoured not only polluting energy resources for their own power, but, at the same time, reinforced social and spatial injustices of oil oligarchy to sustain power (Szeman and Boyer 2017, 9). The connection among patriarchy, petromodernity, and ecological crises under the wide net of the Anthropocene/Andropocene underlies Kay's poem 'Extinction' because upholding patriarchy counteracts sustainable environmental and social futures.

Anthropogenic effects on the environments in which we live create injustices to both humans and nonhumans in socially produced spaces. Although scientific research remains essential to understanding and halting the destructive anthropogenic impacts on the Earth, the arts and humanities can offer insight into forces behind change and ways to study the production of narratives about space that might be reproduced in society. Knowledge alone does not alter human behaviour, whereas influencing change through cultural narratives can alter values and perceptions, and consequently human behaviour.

The historical logic of the Anthropocene – what is also considered to be the modern age of environmental crisis, paralleling the project of modernity within industrial and modernised capitalism – follows an extreme phase of energy production, and particularly hydrocarbon cultures and economies dependent upon the polluting fossil fuels of coal, oil, and gas. One issue that emerges from the symbiotic relationship between the environmental and energy humanities outlined so far in this chapter is that energy functions as a spatial issue as much as a historical and social concern.

Spatialising petromodernity

Neoliberalism has served as the contemporary mode of global capitalism since the 1970s, but with origins going back as far as the late 1940s (Monbiot 2016a, 218). Many scholars over the years have thoroughly defined the various phases of capitalism. The sociologist John Urry, who examines the social dimensions of capitalism and climate connected to 'resources', posits three phases: liberal (mid-nineteenth to mid-twentieth century), nationally organised (1945–1980), and disorganised global capitalism, which is also called neoliberalism (1980–present). Urry is not distinct in framing neoliberalism as a late form of capitalism (Hobsbawm 1994; Mandel 1994), but he isolates the environmental and economic disasters related to the complete reliance on oil and gas as fundamental to this period (2011, 49). Fredric Jameson calls this period of neoliberalism 'multinational or late capitalism', which is the last of three successive historical and spatial versions, starting with 'market capitalism' and moving to 'monopoly capitalism or the age of imperialism'. Jameson associates 'late capitalism' with postmodernism and globalisation (1991, 410), which dovetails with the notion of petromodernity.

Even though Urry outlines social dimensions of the historical logic of capitalism, he is mentioned here because of how he structures capitalism within a resource-based framework, and particularly resources of energy. The current but dwindling phase called 'resource capitalism' overlaps with neoliberalism, which creates various forms of catastrophe caused by economic instability, extreme weather events, and limited resources for human survival (Urry 2011, 49). Although all phases of capitalism rely on 'resources' with no regard for 'nature' as anything but a function of industrial growth, neoliberalism's penchant for oil and gas exploration, production, and addiction, combined with inequitable social and spatial practices, is what has advanced into a contemporary petromodernity from the 1970s to the present.

The world's carbon dependency from the global to the local results from neoliberal capitalism and underlies themes in each of the literary and visual texts discussed in the following three chapters. My aim in this section is to identify the link between fossil fuels and space that produce injustices as a way of conceiving petromodernity as a critically spatial and historical guide for the next three chapters. The environmental problems associated with solastalgia and placelessness or 'no-places' largely emerge from neoliberal forms of resource capitalism, which ultimately create a process of place-taking.

As a theory of political economy, neoliberalism attempts to define humans by market economics. In particular, it proposes societies are best served by the advancement of individual entrepreneurial freedoms and skills in a system that privileges private ownership (particularly of property), free trade, and free markets with little to no state interference. In fact, the role of the state provides a structure for these 'free' practices to flourish in the market, which will maximise the social good (Stiglitz 2002; Harvey 2005a, 2). While shifting from a century of manufacturing, to a consumption and communication capitalist economy known as a new kind of information society (also dependent upon oil), neoliberalism contradicts social forms of justice by promoting competition, increasing tax cuts for the wealthy, deregulating industry, privatising social support services, outsourcing jobs, and eliminating trade unions and collective bargaining.

All of these practices aim to redistribute wealth upward and ultimately restore class power back to pre-twentieth-century levels. In *Spaces of Global Capitalism* (2006), David Harvey succinctly defines neoliberalism as the 'restoration of class power' (9). Neoliberalism serves as an ideological framework that undermines democracy and social justice because it places the power of democratised social institutions and political power into the hands of private interests, not the people, thereby exempting billionaires and larger corporations from paying a large portion of taxes, following environmental laws, adhering to public health and safety, paying workers liveable wages, and being held accountable to the people to which they most affect. Put another way, neoliberal capitalism is a pyramid scheme. It is a private welfare system for the rich funded by the public purse and burdened on the working- and middle-classes. Or, as the geographer Simon Springer upholds, neoliberalism is a 'discourse' that has 'been very successful in convincing us that we should play its zero-sum game', where corporations must hold power at the expense of people and equitable governing social systems (2016, 2).

As a global economic system, neoliberalism relies almost entirely on cheap and available forms of petroleum to function. With new communication technologies and cheap oil (almost free in North America for much of the twentieth century at the expense of the Global South), the global political economy employed what Harvey has famously called a 'spatial fix'. This spatial effect aimed to restore profit for capital by relocating economic centres from high- to low-cost locations, creating a reorganisation of spatial forces in the development (and survival) of capitalism, and particularly where environmental and labour regulations would not compromise efficiency and profit margins (Harvey, 1981, 1; North 2010, 585; Soja, 2010, 90). Spatial forces redirecting economic movement need large amounts of cheap fuel. The neoliberal forms of resource capitalism directly relate to catastrophic global warming in the twenty-first century, as well as unjust social circumstances for many global communities, because of its reliance on fossil fuels.

Spatialising the neoliberal economy underscores what Harvey has also called a process of 'time-space compression', where relationships between various cycles of capitalism manipulate space and time (1990, 240). The relationship between time and space, one that speeds up time while shrinking space, has influenced other social theorists to examine how neoliberalism remains a spatial issue related to social injustice and environmental damage.[5] Harvey elsewhere calls this practice 'accumulation by dispossession', which is the process of the 'privatization of land and the forceful expulsion of peasant populations' (2005b, 159). Neoliberalism can only survive through accumulation by dispossessing people's land or rejecting other human rights. This practice, as we see in the literary and visual texts examined in the following chapters, is an issue of spatial injustice and largely connected to fossil fuel production.

Obtaining natural resources such as oil and gas function as a social production of space, which is to say 'resources' are anything but 'natural' and they are socially produced in the narratives of resource capitalism as essential sources of freedom (Urry 2013, 38). For example, the notion that the Earth provides natural resources for economic development with no other environmental costs underscores the aim of industrial capitalism that produces forms of uneven economic and social development related to oil and gas production. Such an uneven temporal and spatial compression as a result of neoliberal policies creates both environmental and social injustices in three broader ways: shifting economic global centres, moving manufacturing to further markets, and compressing time and space for global production, thereby providing surplus for smaller groups of people. The unclaimed social and environmental price of these three examples, drawing on the work of Harvey, Urry, Marx, and others, are strikingly apparent and worth expanding upon as unjust circumstances of petrospaces.

First, shifting economic energy centres to different global markets increases a need for oil and gas supply and production. The twentieth- and twenty-first-century global economy largely based upon 'cheap' fossil fuels relies on accessible forms of oil and gas transport, such as planes, trucks, trains, and pipelines. The speed of transport changed in the nineteenth century with an energy shift from water to coal (thermal to mechanical) with the invention of the steam engine in 1784 and subsequent expansion of ships and trains. The steam engine powered coal for mining exploration, but then expanded to fuel railroads that transported textile, timber, and metallurgical industries. Energy becomes more mobile with coal and even more so through oil, moving away from geographic constraints of transport, allowing industrialisation exponential expansion around the world. Reinforced perceptions of infinite space equated to endless amounts of energy resources from the Earth.

This compression of time and space within the energy economy of fossil capitalism had obvious environmental effects even at the time of its inception. As Karl Marx noted, extreme uses of 'nature', such as deforestation and coal extraction (and later oil production), resulted in capitalism's 'rift' with nature – a process that robbed both labourers and the soil of any worth. The source of wealth in the fossil economy, moving from coal to oil capitalism between the nineteenth and twentieth centuries, relies upon exploitive precedents of both 'the soil and the worker' (i.e. environmental and social), neither of which has any surplus value (Marx 1976, 637–8; Debeir et al. 1991, 7; Urry 2013, 37). Thus, the exploitation of 'nature', not only labour, created the conditions for industrial capitalism and subsequent forms of neoliberalism later reliant upon communication industries. Marx argued that a capitalist mode of production 'presupposes the domination of man over Nature', which is ecologically and socially unsustainable. He presciently stated:

> All progress in capitalist agriculture is a progress in the arts, not only of robbing the worker, but robbing the soil; all progress in increasing the fertility of the soil for a given time is progress towards ruing the long-lasting sources of that fertility. (1976, 637–8)

As determined in the following chapters, multinational petrochemical corporate practices (as well as agricultural) continue to apply such principles, sacrificing long-term ecological balance for immediate profit. But, they do it in more discrete or 'anonymous' ways. 'The more a country proceeds from large-scale industry as the background of its development', Marx argues, 'the more rapid is this process of destruction' (1976, 637–8).

Mechanical power was considered a liberation from 'nature' in the nineteenth century (rooted in Enlightenment principles of 'improvement' and eventually powered by the steam engine). No longer considered a liberator, but the genesis of mass global pollution, industrialised mechanical power continues to haunt society in the twenty-first century in a what we might now call a return of the ecologically repressed within the Anthropocene. Returning to Buell's apt linkage, the exuberance of the fossil fuel economy literally fuelled environmental catastrophe (2012, 276). The systematic degradation of ecosystems and species around the world, what Marx referred to as 'nature', proves to be the greatest challenge for survival. These seeds were laid within the capitalist modes of production that now use shifting global markets as ways to exploit the 'soil and the worker' through a compression of time and space.

Second, the movement of manufacturing and the subsequent transportation of goods to further markets increase energy use and emissions more than reducing them (North 2010, 585). This system of colonisation originated in the 1400s during which time capitalism expanded to global markets because of imperial exploration and domination. Crops such as coffee, tea, spices, and sugar raised in the Americas, Africa, and Asia were then exported back to Europe. This structure has only expanded in complexity over the past several centuries, with food markets still rooted in the imperial 'peripheries' of the Americas (South, Central, and North) and exported to industrialised countries, but, as established in the following chapters, also continued to oppress already marginalised populations to extract oil and gas.

Third, compressing time and space for global production to increase capital for a small few who own the majority of the wealth creates an unjust economic system. It unevenly privileges the owners of capital over the global population, thereby perpetuating an unsustainable system leading to extreme forms of income inequality and mass poverty juxtaposed against obscene wealth. It also disenfranchises populations by shifting jobs to markets with little to no regulation of environmental (and labour) protection.

One international example of this is in Haiti. The impoverished and undernourished Indigenous populations of Haiti struggle to survive growing food on deforested and eroded mountain slopes, while multinational agribusinesses control the fertile land in the valleys, producing masses of food for global markets. Haiti also serves as an example of the time-space compression of capital-intensive and energy-intensive hybrid seeds, chemical fertilisers, pesticides, and farm machinery used by agribusiness (Foster 1999, 22). Genetically modified organisms (GMOs) largely dominate in these 'further markets' around the planet because of deregulated spaces, not too dissimilar than the exploitive use of so-called 'open spaces' in the height of European colonisation in the sixteenth and seventeenth centuries. This rapid growth of productivity reliant on cheap forms of fossil energy constructs systems of spatial injustice for local populations, while also expanding unsustainable systems of economic and environmental growth.

Another example of spatial injustice tied to global production is the financial crash in October 2008. Soja's book *Seeking Spatial Justice* (2010), and the motivation to reintroduce space as a phenomenon of social injustice, was partly inspired by examining the spatial effects of the global financial crisis. Gillian Tett reduces the cause to several individuals connected with J.P. Morgan who invented the credit default swap in the mid 1990s, which ultimately produced the 2008 crash. Gillian's book *Fool's Gold* (2009) outlines the 15-year history of high-risk and cavalier banking practices, which epitomise neoliberalism's penchant for deregulation. The credit default swap is a complex type of credit derivative that permits banks to 'swap' loan risk to outside investors; it allows a bank to then free up more credit to other businesses. Default risk is a common aspect of banking, where banks judiciously lend money to people while keeping capital aside for potential losses, but credit derivatives are not. 'Risk' could be turned into a financial product and sold, similar to shares, bonds, or oil. The only problem is that capital did not exist for losses, hence the 'bubble' concept.

It is not surprising this volatile and circular system of lending came crashing down on everyone in the global financial system. While the banks were mostly bailed out, most people were not, especially working-class families who held mortgages on homes based upon non-existent credit. Gillian's analysis concludes that the 'Morgan Mafia' created a self-inflicted 'catastrophe' costing the world between $2,000 and $4,000 billion, while also increasing wealth at the top of the spectrum. The credit default swap shaped a non-linear system of finance that affected people out of space and time of this immediate group. A decision made by some clusters of people in New York contained catastrophic effects, compressing space and time, causing massive global injustice that remains unrectified (Foley 2009; Tett 2009). Volatile oil prices linked to credit derivative lending and provoked by oil shortages and price increases in the mid 2000s also contributed to the crash (Urry 2013, 191). A similar spatially unjust scenario could be argued about climate change, where small groups of entitled individuals hold the power to make decisions that influence the potential survival of human and nonhuman species because of the web of interconnected systems (see Chapter 7).

These three examples of space-time compression resulting from capitalism illustrate that the construction of petrospaces produces many forms of spatial injustice. The petroleum industry has mostly propagated (and funded) neoliberalism in the post-World War II Western world (Urry 2011, 117–21; Huber 2013). In resource capitalism, the cost of resource extraction and it subsequent price of pollution, displacement, and natural disasters is not configured into the economic analysis because resources are considered free. Urry specifically notes the Great Crash of 2008 challenged the notion that resource capitalism of the twentieth century can economically or socially endure long term. As British environmental writer George Monbiot points out, neoliberalism is 'the God that failed' (2016b). What is left once people accept that neoliberalism has failed? The neoliberal project only deepens the more we remain in crisis because it produces these crises, whether it is 'war' (on drugs, terrorism, immigration, etc.) or economic austerity. As the capitalist saying goes, 'never let a good crisis go to waste'. What happens to the energy that fuels the neoliberal project? What happens when oil disappears or destroys our capacity to live on this planet (or whichever comes first)? Monbiot notes that 'the zombie doctrine [of neoliberalism] staggers on, and one reason is its anonymity' from society (2016b). Oil fuels this zombie system, ironically providing energy to a dead economic body limping forward.

Oil and gas extraction and production, as symptoms and functions of neoliberalism, have been somewhat shrouded in this public anonymity, largely out of sight and sequestered in remote regions of the Earth, unlike the centralised urban smokestacks spewing out particulates from burning coal during the Industrial Revolution. While the images of an offshore oil platform or oil derrick often appear in seemingly remote spaces, such as in the middle of an ocean or in rural Texas, it does not have the same direct visual and therefore emotional impact as a smokestack in East London or in Pittsburgh, Pennsylvania. In *Carbon Nation* (2014), historian Bob Johnson acknowledges this 'anonymity' of oil culture: 'we industrial peoples have preferred to keep our energy dependencies out of sight' (xxix). While oil production associated with the Texas oil derricks was a source of pride in the 1970s, they have become increasingly hidden in a time of public awareness of the climate crisis and its link to fossil fuels. Oil once was fetishised openly, but it has become less visible in society and culture both physically and perceptually (Macdonald 2017, 38). Increasing urban density in the twenty-first century also distances visual extraction of fossil fuels or minerals from rural spaces that fuel society from plain sight. Such rural spaces of extraction are now heavily guarded, blocking public access from reporting or any forms of documentation and video recording.

The hidden aspects of energy dependency and production also include public discourses about them. When the effects of oil fuelling the zombie system of neoliberalism become transparent, they become more difficult for society to support socially and politically. Neoliberalism (and the oil economy driving it often referred to as 'extractive capitalism') resists being named or located in time and space. Critical approaches in the energy humanities, and particularly petrocultures focused on oil and gas, illuminate the zombie system of oil and how it staggers on despite the destruction it reaps. Allowing an economic system driven by polluting energy systems to be unseen permits the controllers and owners of that system to have ultimate anonymity and freedom of movement without public resistance. Eventually, zombie oil becomes visible in our daily lives as long as we know what to look for and the spaces in which it exists.

Place-taking

Drawing on these historical lineaments of petromodernity viewed also as a petro-space, it is evident that neoliberalism is seemingly placeless, producing an effect of placelessness tied to spatial injustice and solastalgia. An overabundance of commodification and commerce is not connected to any sense of place, but rather to a non-place. 'Place', maintains Tim Cresswell, 'has also become a political symbol for those who want to fight against the ever-present power of global capitalism' (2015, 95). A non-sustainable relationship with environments on the Earth, reducing people's sense of place, will not only affect the ability of the Earth to sustain life. It will also alter the emotional and psychic health of humans and non-humans experiencing a loss of place through preventable ecological damage. In *Justice, Nature, and the Geography of Difference* (1996), Harvey cites an article by Kirkpatrick Sale in *The Nation* that confirms the need for a place-based consciousness to combat catastrophic ecological change: 'The only political vision that offers any hope of salvation is one based on an understanding of, a rootedness in, a deep commitment to, and a resacralization of place' (302).

The spatial containers of global neoliberalism are what anthropologist Marc Augé refer to as 'non-places' of 'supermodernity', which generate places of 'triple decentring' that include 'spaces of circulation, consumption and communication', such as airports, borders, supermarkets, shopping malls, and hotel chains (or any homogenising corporate chain). Such non-places are formed in relation to certain ends (transport, transit, commerce, leisure) and 'they will always be political' (Augé 2008, 92). They result from supermodernity, which is a commodity-based culture devoid of humanity and lacking any sense of place or personal connection. Non-places are bereft of communal relations, producing individuals void of identity with no relational (spatial) or temporal (historical) relationship to place. People are situated in a world of supermodernity (*surmodernité*), which for Augé signifies a space lacking consciousness (thinking back to Monbiot's zombies), representing an end of modernity (or an illusion to the end of 'progress'). Ultimately, supermodernity delocalises and accelerates spaces for consumption (Augé 2008, 21–2). Opposite to the concept of social and environmental connection of place-home outlined in Chapter 2, non-places create disconnection and fragmentation.

Rather than cultivating a process of place-making that enhances social and human connection, neoliberalism produces non-places of place-taking, leaving humans and nonhumans without an emotional correlation to place (i.e. place-home) or, in some cases, the resources and environmental health to survive. Place-taking can be real, where freedoms of mobility are reduced, as well as conceptual through limiting ideas, expression, or ways of being in one's place. It can also be residual through extractive practices that colonise spaces for resources, only leaving waste behind. The Anthropocene is a period largely producing non-places, where power is created through space used for extracting resources or ideas with displacing and disorientating consequences for humans and nonhumans, instead of producing space that generates sustainable or egalitarian societies.

The process of place-taking resulting from compressed space and time in neoliberalism creates solastalgia. As with the psychic affects of solastalgia, generating feelings of homesickness or exile caused by environmental damage in places people live, people cannot establish place-home in these ambiguous and fragmented containers of space that are constantly altering local environments. People feel eco-anxiety about their environments and ways of being in the world. In the non-places of supermodernity, which is largely the effect of neoliberalism and extractive practices of petromodernity, people never feel at 'home'. Place-home is literally being affected by extraction (gas pipelines, fracking, mining, offshore oil) or the spatial consequences of it elsewhere on the Earth through changing climates (species extinction, water and air pollution, ocean acidification, rising sea levels). The anonymity of place allows for a 'zombie' population to ignore the failing system of petrocapitalism. Unmasking a system that supports vast inequality and suffering, with massive levels of global poverty not witnessed since the nineteenth century, increases the possibility for public response to end such a system.

One lever of such change revealed throughout this book is through artistic creation in the humanities. As a mode of production that historically documents and at times confronts human and nonhuman suffering, literary and visual culture draws on universal themes, story, and images that affect values and perceptions about how we want to change society to one of place-making instead of place-taking. The response to simultaneities of 'continuity and discontinuity, local and global, place and non-place', for Augé, 'emerges in art and artistic creation' (2008, xvii). He notes that 'today's artists and writers' might be 'doomed to seek beauty in "non-places"', but this engagement does not produce an inferior aesthetic as it might intimate (Augé 2008, xxii). Rather, some artists and writers, whether consciously or not, highlight non-places that are socially produced by capitalist systems to address effects of solastalgia as a form of spatial injustice, where people once had connection to a centralised place, but are now more fragmented and displaced.

Whether it is the Oil Clearances of the Highlands portrayed in drama (Chapter 4), the production of oil and gas in Shetland displayed in a filmpoem (Chapter 5), or pipeline construction off the west coast of Ireland captured in a documentary film (Chapter 6), the literary, performative, and visual texts outlined in subsequent chapters of Part II address the loss of place-home while also reclaiming it in the social and cultural narrative. Artistic production in the humanities motivated by social and political change (or outrage or critique) make sense of simultaneities of 'continuity and discontinuity' in time-space compression for audiences, particularly while living in a quickening and decentralised capitalist world-system.

Culture and oil

Drawing on petrocultures within the Anthropocene, and amid the space-time compression of neoliberalism and its place-taking consequences, this section finalises the links between oil and culture. Oil viewed as a contemporary petroculture proves to be a multidimensional topic in the humanities with many spatial properties. It is concurrently real and imagined, visual and tactile, personal and public, and dangerous and beneficial. We all have a personal and subjective relationship to oil as much as we do to the built and non-built environments that surround us.

This extreme subjectivity, what Brett Bloom has called 'petro-subjectivities' in his book about 'de-industrialising' our 'sense of self', is something that remains ubiquitous but often unacknowledged. As Bloom argues, 'Oil produces our daily lives, our daily selves, our daily communities and everything else in a primary way' (2015, 19). Bloom's petro-subjectivity map illustrates the interconnected system of oil in our lives, highlighting links among culture, economics, health, food, and transportation across time and space (see Figure 3.2). For my purposes here, Bloom's oil map provides a visual of petrospaces, concretising the components discussed in this chapter. The petromodern energy history associated with fossil fuels remains intertwined with conceptualisations, symbols, and representations related to bodies, psychic identity, social structures, technologies, and environments in material and symbolic cultures (Buell 2012, 273). Because oil is quotidian, despite its superficial anonymity from public view, it is not surprising that it conceptually and physically shapes global economics, culture, and politics. Environmental writers, artists, and activists have responded to the culture of oil as its own critical system to magnify its addictive presence in our lives.

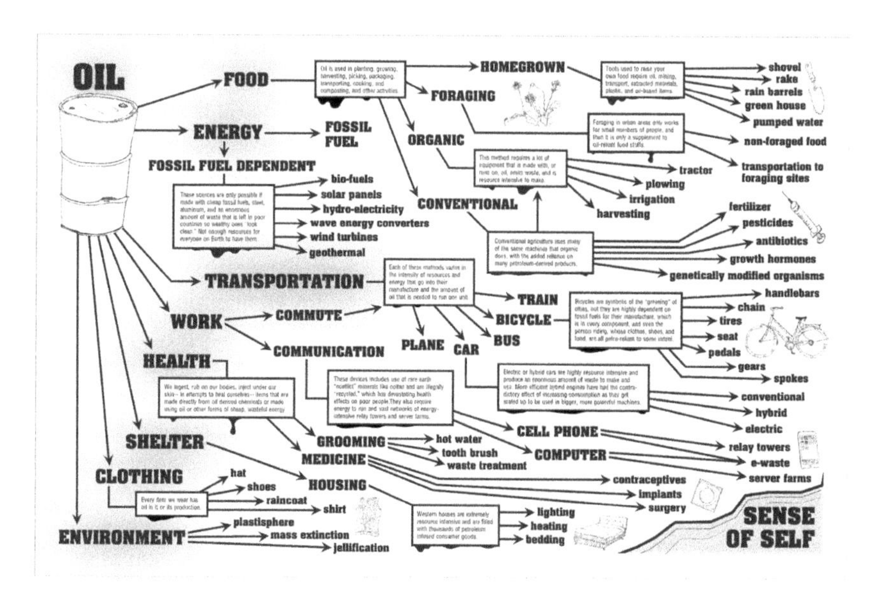

Figure 3.2 'Petro-Subjectivity Map' (Brett Bloom with illustrations by Bonnie Fortune, 2015), with kind permissions from Bloom.

Szeman has noted that oil can be viewed simultaneously as a problem and possibility, which implies reciprocity to other social and cultural narratives within the energy humanities (2013, 146). Oil creates forms of spatial injustice around the planet, allowing for expansive lifestyles in some parts but not in others. One example would be the amount of energy use in the Global North (industrialised areas in most cases) compared to the Global South. The average North American uses 40 times the amount of commercial energy than an average person in developing countries. A more extreme example is that of Sub-Saharan Africa, where people consume 80 times *less* energy than the average person in an advanced capitalist country (Foster 1999, 19). The ongoing quest for oil is geopolitically imbalanced, favouring certain markets over others, which creates concealed zones of exploitation, as well as unsustainable social and environmental costs (Macdonald 2015, 55).

Oil is not only physical and material, but also a socially produced idea, animating various abstract categories such as justice, freedom, or oppression (Szeman 2013, 146). Szeman acknowledges there exists a 'social ontology of oil', which explains 'the how, why and wherefore of oil in our social, cultural and political life' (2012, 426). Adding to Szeman's social ontology of oil, we can also include space as an integral aspect of understanding how oil functions in our lives. Oil, for some equates to freedom when governments justify tax breaks for petrochemical companies. They would argue that allowing cheaper forms of oil creates necessary lifestyles for Western society. Cheap oil equals to a form of freedom for Western societies dependent upon it. This is clearly a socially constructed idea of space because 'freedom' in one part of the Earth, such as the Global North, comes with social and environmental consequences in other parts, as in the Global South. This latter scenario equates oil with oppression. While oil derricks in Iraq might signify independence, freedom, and wealth for some countries, they also signal instability and tyranny for others.

Because oil shapes social and human history, it also shapes the history of the humanities and cultural critique within it. Macdonald argues:

> Registering oil's sheer significance in modern petro-life – myriad material, representational, and non-manifest forms – provides a platform for interpretive and imaginative disciplines to elicit new ways to consider its past, present, and future. It also presents a challenge for Humanities scholars and cultural practitioners to make good on the claim that how we read, cognize, visualize, narrate, perform, and represent oil is connected to the social and cultural way we inhabit and are habituated to it. (2017, 36)

For scholars such as Macdonald, among others discussed in this chapter, contemporary culture is a petroculture, where fossil fuels have shaped the character and form of the modern (Buell 2012; LeMenager 2012; Wilson and Pendakis 2012; Barrett and Worden 2014; Szeman and Boyer 2017). Since oil encompasses the objects we use, buildings we occupy, or the spaces we move through in our daily routines, it also is a social and spatial human question, even though framing oil in what we can acknowledge as a petroculture has advanced little to this point outside of the energy humanities (Szeman et al. 2016, 14, 17).

As this chapter attempts to show, oil is not only a source of energy, but also a spatial and environmental concern in the ways it has shaped contemporary culture and society, and the artistic production that responds to it. Our dependence on oil creates highly spatial conceptual and physical relationships between language and society within our everyday lives. This is partly why the *LEGO: Everything Is NOT Awesome* viral YouTube video discussed in the previous chapter attracted such attention and mobilised action. It explicitly displays the interconnections among politics, corporate practices, and oil. As the video argues, these unjust practices directly associate with children's imaginations, or 'unspoilt wilderness', and how they develop. That is to say, oil shapes us, informs us, and defines our very selves within a history of modernity – or, what we might more accurately call a petromodernity.

Transitioning from a social and spatial critique in this chapter to literary and visual analysis in the following three chapters, I want to conclude somewhat unconventionally with another 'climate change poem for today'. 'Silent Sea' by the British poet Rachael Boast amplifies 'our foolish husbandry' that lies 'under oil'. It is about the 'silent sea' once used as a phrase by the Romantic poet Samuel Taylor Coleridge, who was at times responding to the beginnings of industrial capitalism. The speaker in 'Silent Sea' experiences the 'vessel' of industry and its polluting effects both socially and environmentally. Under this vessel, 'Nothing can breathe / under oil, nor register that / dark membrane's slick / over sight'. The poem goes on to say, 'a metallurgy' full of 'false gold' is 'too often' of 'talk, and talk and talk'. The 'vessel' of petromodernity is ultimately 'manned' (literally by patriarchy) 'by non-thinking from non-feeling' and says 'nothing at all' (Boast 2015).

The sea remains 'silent' amid the industrial history of resource capitalism fuelled by fossil fuels. Oil remains a twentieth- and twenty-first-century concern, despite developments in more sustainable forms of energy use, and so the following three chapters in Part II will look at its effect upon contemporary culture. The petrocultures pervading peoples' lives motivate literary and visual texts as creative responses rather than remaining 'silent' in reaction to spatial injustices.

Notes

1 See stated aims on the Petrocultures website: http://petrocultures.com/. See also Wilson et al. (2017).
2 See the LAGI website: http://landartgenerator.org/index.html.
3 Beginning in 2010, LAGI is an art-architecture competition that has held biennial competitions in Dubai Abu Dhabi (2010), New York City (2012), Copenhagen (2014), and Santa Monica (2016), and Melbourne (2018). See the competitions at http://landartgenerator.org/competition2014.html.
4 See the website for Alien Energy: http://alienenergy.dk. Alien Energy is also part of the LAGI.

5 Smith's study (2010 [1984]) on uneven development serves as almost a companion volume to Harvey (1990). The postcolonial theorist Edward Said claimed Smith's groundbreaking study is 'a brilliant formulation of how the production of a particular kind of nature and space under historical capitalism is essential to the unequal development of landscape that integrates poverty with wealth, industrial urbanization with agricultural diminishment' (1993, 225).

Bibliography

Augé, Marc. 2008 [1995]. *Non-Places: An Introduction to Supermodernity*. London: Verso.

Barrett, Ross, and Daniel Worden. 2014. 'Introduction'. In *Oil Culture*, edited by Ross Barrett and Daniel Worden, xvii–xxxiii. Minneapolis, MN: University of Minnesota Press.

Bateson, Gregory. 1972. *Steps to an Ecology of Mind: Collected Essays in Anthropology, Psychiatry, Evolution, and Epistemology*. Chicago, IL: University of Chicago.

Berman, Marshall. 1982. *All That Is Solid Melts into Air: The Experience of Modernity*. New York: Penguin.

Bloom, Brett. 2015. *Petro-Subjectivity: De-Industrializing Our Sense of Self*. Ft. Wayne, IN: Breakdown Press.

Boast, Rachael. 2015. 'A Climate Change Poem for Today: Silent Sea', *The Guardian*, 29 May. Accessed 14 November 2016. www.theguardian.com/environment/2015/may/29/a-climate-change-poem-for-today-silent-sea-by-rachael-boast.

Boyer, Dominic, and Imre Szeman. 2014. 'Breaking the Impasse: The Rise of the Energy Humanities'. *University Affairs* 40, 12 February. Accessed 2 May 2015. www.universityaffairs.ca/opinion/in-my-opinion/the-rise-of-energy-humanities/.

Buell, Frederick. 2012. 'A Short History of Oil Cultures: Or, the Marriage of Catastrophe and Exuberance'. *Journal of American Studies* 46(2): 273–93.

Chakrabarty, Dipesh. 2009. 'The Climate of History: Four Theses'. *Critical Inquiry* 35: 197–222.

Cresswell, Tim. 2015 [2004]. *Place: An Introduction*. Malden, MA: Wiley Blackwell.

Crutzen, Paul J. 2002. 'Geology of Mankind'. *Nature* 415, 3 January: 23.

Crutzen, Paul J., and Eugene Stoermer. 2000. 'The Anthropocene'. *Global Change Newsletter* 41: 17–18.

Davis, Jeremy. 2016. *The Birth of the Anthropocene*. Berkeley, CA: University of California Press.

Debeir, Jean-Claude, Jean-Paul Deléage, and Daniel Hémery. 1991. *In the Servitude of Power: Energy and Civilization through the Ages*. Translated by John Barzman. London: Zed Books.

Farrier, David. 2016. 'How the Concept of Deep Time Is Changing'. *The Atlantic*, 31 October. Accessed 30 November 2016. www.theatlantic.com/science/archive/2016/10/aeon-deep-time/505922.

Foley, Stephen. 2009. '*Fool's Gold*, by Gillian Tett: How the Geeks Broke the Banks'. *The Independent*, 30 April. Accessed 18 January 2017. www.independent.co.uk/arts-entertainment/books/reviews/fools-gold-by-gillian-tett-1676686.html.

Foster, John Bellamy. 1999. *The Vulnerable Planet: A Short Economic History of the Environment*. New York: Monthly Review Press.

Ghosh, Amitav. 1992. 'Petrofiction'. *New Republic* 2: 29–34.

Harvey, David. 1981. 'The Spatial Fix: Hegel, von Thünen and Marx'. *Antipode* 13(3): 1–12.

——. 1990. *The Condition of Postmodernity: An Enquiry into the Origins of Cultural Change*. Oxford: Blackwell.
——. 1996. *Justice, Nature, and the Geography of Difference*. Cambridge, MA: Blackwell.
——. 2005a. *A Brief History of Neoliberalism*. Oxford: Oxford University Press.
——. 2005b. *The New Imperialism*. Oxford: Oxford University Press.
——. 2006. *Spaces of Global Capitalism: Towards a Theory of Uneven Geographical Development*. London: Verso.
Hobsbawm, Eric. 1994. *The Age of Extremes: A History of the World, 1914–1991*. New York: Vintage.
Huber, Mathew. 2013. *Lifeblood: Oil, Freedom and the Forces of Capital*. Minneapolis, MN: University of Minnesota Press.
Johnson, Bob. 2014. *Carbon Nation: Fossil Fuels in the Making of American Culture*. Lawrence, KS: University Press of Kansas.
Jameson, Fredric. 1991. *Postmodernism: Or, the Cultural Logic of Late Capitalism*. Durham, NC: Duke University Press.
Kay, Jackie. 2015. 'A Climate Change Poem for Today: Extinction', *The Guardian*, 15 May. Accessed 14 November 2016. www.theguardian.com/environment/2015/may/15/a-climate-change-poem-for-today-extinction-by-jackie-kay.
Lakoff, George, and Mark Johnson. 1980. *Metaphors We Live By*. Chicago, IL: University of Chicago Press.
LeMenager, Stephanie. 2012. 'The Aesthetics of Petroleum, after Oil!' *American Literary History* 24(1): 59–86.
——. 2014. *Living Oil: Petroleum Culture in the American Century*. New York: Oxford University Press.
Macdonald, Graeme. 2015. 'Commentary'. In *The Cheviot, the Stage, and the Black, Black Oil* by John McGrath, edited by Graeme Macdonald, 17–80. London: Bloomsbury.
——. 2017. 'Containing Oil: The Pipeline in Petroculture'. In *Petrocultures: Oil, Energy, Culture*, edited by Sheena Wilson, Adam Carlson, and Imre Szeman, 36–77. Montreal-Kingston: McGill-Queens University Press.
Malm, Andreas. 2016. *Fossil Capital: The Rise of Steam Power and the Roots of Global Warming*. London: Verso.
Mandel, Ernest. 1994 [1975]. *Late Capitalism*. London: Verso.
Marx, Karl. 1976. *Capital*, vol. I. New York: Vintage.
McNeill, John R., and Peter Engelke. 2014. *The Great Acceleration: An Environmental History of the Anthropocene since 1945*. Cambridge, MA: Belknap Press.
Mitchell, Timothy. 2011. *Carbon Democracy: Political Power in the Age of Oil*. London: Verso.
Monbiot, George. 2016a. *How Did We Get into the Mess?* London: Verso.
——. 2016b. 'Neoliberalism: The Ideology at the Root of All Our Problems'. *The Guardian*, 15 April. Accessed 25 October 2016. www.theguardian.com/books/2016/apr/15/neoliberalism-ideology-problem-george-monbiot.
Moore, Jason. 2015. *Capitalism in the Web of Life: Ecology and the Accumulation of Capital*. London: Verso.
——, ed. 2016. *Anthopocene or Capitalocene? Nature, History, and the Crisis of Capitalism*. Oakland, CA: PM Press.
Morton, Timothy. 2013. *Hyperobjects: Philosophy and Ecology after the End of the World*. Minneapolis, MN: University of Minnesota Press.
North, Peter. 2010. 'Eco-Localisation as a Progressive Response to Peak Oil and Climate Change: A Sympathetic Critique'. *Geoforum* 41: 585–94.

Said, Edward. 1993. *Culture and Imperialism*. New York: Vintage.
Smith, Neil. 2010 [1984]. *Uneven Development: Nature, Capital and the Production of Space*. London: Verso.
Soja, Edward. 2010. *Seeking Spatial Justice*. Minneapolis, MN: University of Minnesota Press.
Springer, Simon. 2016. *The Discourse of Neoliberalism: An Anatomy of a Powerful Idea*. London: Rowman & Littlefield.
Stiglitz, Joseph. 2002. *Globalisation and Its Discontents*. London: Allen Lane.
Szeman, Imre. 2012. 'Crude Aesthetics: The Politics of Oil Documentaries'. *Journal of American Studies* 46(2): 423–39.
——. 2013. 'How to Know about Oil: Energy Epistemologies and Political Futures'. *Journal of Canadian Studies* 47(3): 145–68.
——, and Dominic Boyer. 2017. 'Introduction: On the Energy Humanities'. In *Energy Humanities: An Anthology*, edited by Imre Szeman and Dominic Boyer, 1–13. Baltimore, MD: Johns Hopkins University Press.
——, and Jeff Diamanti. 2017. 'Beyond Petroculture: Strategies for a Left Energy Transition'. *Canadian Dimension* 51(1). Accessed 28 March 2017. https://canadiandimension.com/articles/view/beyond-petroculture-strategies-for-a-left-energy-transition.
Szeman, Imre, and the Petrocultures Group. 2016. *After Oil*. Morgantown, WV: West Virginia University Press.
Tett, Gillian. 2009. *Fool's Gold: How Unrestrained Greed Corrupted a Dream, Shattered Global Markets and Unleashed a Catastrophe*. London: Little, Brown.
Urry, John. 2011. *Climate Change & Society*. Cambridge: Polity Press.
——. 2013. *Societies Beyond Oil: Oil Dregs and Social Futures*. London: Zed Books.
Whitney, Rob. 2013. 'The Future Is Looking Musical: WEC Scenarios to 2050'. In *World Energy Insight 2013*, edited by David Buchan, 16–17. London: First.
Wilson, Sheena, and Andrew Pendakis, eds. 2012. 'Sighting Oil'. *Imaginations: Journal of Cross-Cultural Image Studies* 3(2). Accessed 9 February 2017. http://imaginations.csj.ualberta.ca/?p=3869.
Wilson, Sheena, Imre Szeman, and Adam Carlson. 2017. 'On Petrocultures: Or, Why We Need to Understand Oil to Understand Everything Else'. In *Petrocultures: Oil, Energy, and Culture*, edited by Wilson, Sheena, Imre Szeman, and Adam Carlson, 3–19. Montreal and Kingston: McGill-Queen's University Press.

4 Speed of petrodrama

North Sea oil and Britain

Oil was first discovered in the North Sea in December 1969. Although earlier North Sea explorations of gas had produced varying results throughout the 1960s, Phillips Petroleum first unearthed oil in 1969 around Norwegian waters at Ekofisk. A month later, a drilling-rig called the *Sea Quest*, which British Petroleum (BP) owned and chartered to the American Oil Company (Amoco), located oil in the Montrose Field about 217 km north-east of Aberdeen, Scotland. Thus, British North Sea oil was born. The discovery ushered in a new era of economic and political transition that continues to garner cultural and literary responses in the humanities. This chapter examines John McGrath's 1973 play *The Cheviot, the Stage, and the Black, Black Oil* (2015), which is about contested Scottish Highland histories and spaces during the North Sea oil boom. Through the flexible spatial form of 'agitprop' theatre, McGrath's play confronts ecologically and socially damaging oil infrastructures and mobilises action through his own form of petrodrama.

The 'momentous' North Sea oil and gas unearthing changed the economic and social outlook for future decades in Britain (and particularly those in control of Westminster). In 1977, Britain's Prime Minister, James Callaghan of the Labour Party, proclaimed: 'God has given Britain her best opportunity for one hundred years in the shape of North Sea oil' (cited in Harvie 1994, 2). Between 1977 and 1985, Britain profited £100 billion from North Sea oil and gas. But, as substantial economic reliance on fossil fuels has proven in the past for an oil state's mutable economy, Britain's generation of wealth from oil and gas did not solve the national deficit. This issue was partly economic, based upon the fluctuating price of oil in international markets, but it was mostly connected to the politics surrounding redistributing wealth. Conservative practices under Margaret Thatcher's Conservative Party, in particular, siphoned sizeable revenues from North Sea oil and gas to balance a deficit that continued to increase rather than lessen because of an upper-class welfare system providing tax cuts to wealthy individuals and corporations. North Sea oil prices were expected to increase in 1979–1980 from £2.3 billion to £5.3 billion in 1983–1984 (Buiter and Miller 1981, 342; Harvie 1994, 2–3). As the economic historian Sidney Pollard points out, 'Without the oil, the Thatcher experiment would almost certainly have been cut short as early as 1981 or 1982' (cited in Harvie 1994, 286).

Thatcher's ongoing mismanagement of North Sea oil and gas was partly a product of ignorance and partly one of elitism, the latter of which signals an unwillingness to engage with social 'commerce' (of which the feared unions were a part), reproducing a new or neoliberal form of British imperial nationalism. Britain, while 'sliding politely into post-imperial and post-industrial decline, seemed the least likely candidate on earth for membership among the great oil-producing nations', according to the Occidental Petroleum representative Armand Hammer. Thatcher's Conservative government, Hammer continues, responded to the emerging petroeconomy 'carelessly and complacently' and 'practically threw it into the hands of the Seven Sisters [the global oil cartel between the 1940s and 1970s]' (cited in Harvie 1994, 264). Hammer's comments affirm the mismanagement of North Sea oil and the potential economic benefits lost to the people of Britain (and particularly Scotland), despite the rather large profits North Sea oil produced.

North Sea oil and gas serve as Britain's 'third industrial revolution', but it is, as the historian Christopher Harvie observes, a resource-based history that remains more peripheral than previous industrial revolutions (1994, 3). Eric Hobsbawm has pointed out revenues at the peak of North Sea oil production in the 1980s equaled 9 per cent of the total tax revenue. However, financial benefits from North Sea oil did not translate into direct investment in manufacturing and industry to create jobs, especially those jobs left in the vacuum of the mining industry during the 1970s and 1980s (Hobsbawm 1999, 309–10).

The so-called oil revolution has also produced an inordinate amount of creative output in the humanities, speaking back to and about oil in Britain. In particular, a Scottish cultural renaissance emerged in the 1970s and 1980s responding to declining industrialisation, as well as the economic and socially displacing forces of oil. The consequences of deindustrialisation and the multinational priorities of petroleum corporations generated resistance campaigns to unite Scotland with the European Union, rather than as part of Britain, starting in the 1990s and continuing into the present with the 2014 Scottish Independence Referendum.[1] Scotland received much less profit from the North Sea oil and gas boom years, and this was particularly the case during Thatcher's neoliberal policies in the 1980s. Besides becoming a hub for North Sea oil production and transport for decades to come, Scotland's petroculture influenced cultural and social reactions to it.

One of the earliest cultural responses to the discovery of oil was John McGrath's avant-garde play *The Cheviot, The Stag, and the Black, Black Oil.* Similar to other works by McGrath, the play is a social and political statement on Scotland's modern history of colonialism, capitalism, and territorial disenfranchisement. McGrath already had an established record of writing for repertory theatre, as well as film and television, with many credits to his name, including the celebrated 1960s and 1970s BBC police drama set in the north of England titled *Z Cars*. Since graduating from the University of Oxford, his artistic ambition was to combine aesthetics and culture to produce social and political change. McGrath outlines his distinctive writing approach where he discusses how the aim of 'political' theatre comprises three steps:

> firstly, the struggle within the institutions of theatre, against the hegemony of the 'bourgeois' ideology within those institutions; secondly, the making of a theatre that is *interventionist* on a political level, usually outside those institutions; and thirdly and most importantly, the creation of a counter culture based on the working class, which will grow in richness and confidence until it eventually displaces the dominant bourgeois culture of late-capitalism. (1980, 44; original emphasis)

Such views prompted him to establish a politically focused theatre company, which produced *The Cheviot* as its first play.

The Cheviot was part of a larger artistic campaign to challenge social injustices through the art form of theatre. McGrath, along with his partner Elizabeth MacLennan, as well as David MacLennan, founded the Scottish theatre company called 7:84 in 1971. The name derived from a statistic published in *The Economist* in 1966 that stated 84 per cent of the wealth in Britain was owned and controlled by 7 per cent of the population, thus leaving 93 per cent of the population with only 16 per cent of the wealth. Just by virtue of its name, 7:84, the company aims to confront social and economic inequality through a place- and class-based response to unequal ownership of territory, capital, and power (Macdonald 2015, 17). 7:84 draws on a tradition of agitprop theatre (i.e. agitation and propaganda), which originated in Russia as an artistic source of communist propaganda theatre and film, but has now become regarded as a form of political art used to challenge injustices in an age of capitalism.

Alternative British theatre thrived in the first half of the 1970s, with over 150 groups throughout the UK. Although many of these troupes existed for short intervals, few sustained the impact of 7:84. McGrath began the Scottish 7:84 Theatre Company, which differed somewhat from the English 7:84 version, as a collective aimed at promoting Marxist approaches to local history, but particularly through subregional touring shows celebrating Scottish resistance to centuries of repression and exploitation (Kershaw 2002, 138). The goal, as it was for many avant-garde theatre companies in the 1970s, was to erase the boundaries between working-class communities and the formal and more elitist structure of theatre performances, which were historically produced and performed for the upper classes. The approach of 7:84 reached an entirely new class of audience, representing people without voice or legitimacy in British society (Kershaw 2002, 145).

As the first play to be produced, *The Cheviot* remains the most highly regarded of 7:84 performances to date. It premiered on 31 March 1973 at the George Square Theatre in Edinburgh. Remarkably, McGrath assembled the performers only a couple of weeks before and finished the script the night before the performance. Over the next few years, *The Cheviot* performed 100 shows during a 17,000-mile tour; it ultimately reached around 30,000 people in Scotland (Harvie 1994, 271).

Such a vigorous run for an avant-garde theatre production signals the direct impact of live performances. Social theatre addresses prescient political issues with the aim of reaching communities about circumstances affecting them directly. As stated in his three points of political theatre, McGrath's focus could not be clearer: to create 'a counter culture based on the working class'.

The Cheviot contains the three fundamental pieces of a 7:84 production: place, capital, and power. The plot is simple enough, but the execution uses many different styles of theatre, which is complex and difficult to perform. Members of the cast must be multitalented, with musical, comic, improvisational, and dramatic skills. Harvie succinctly summarises *The Cheviot* as a 'history, political cabaret and *ceilidh*' of the Highlands (1994, 271; original emphasis). The play spans over 200 years and centres on three specific historical junctures in Scotland: the Highland Clearances, which exposes early forms of capitalist agribusiness over land rights; imperial tourism of the Victorian and Edwardian 'ruling classes'; and the petroleum economy of the North Sea in the 1970s. While examining these three contested periods in modern Scottish history, the play's multidimensional and multi-genre form parallels historical transitions, which also utilise episodic leaps back and forth in time, by employing musical, pantomime, farce, and tragedy. At the same time, the play uses audience interaction and improvisation, through real and fictional characters, across various territories and geographical spaces of the Highlands.

Trying to decipher a concise overview of the play might not be as beneficial as understanding its context within a social and political history. Baz Kershaw states, 'there is no plot to summarize; instead, there is a compendium of popular theatre forms' (2002, 202). Perhaps the play's so-called 'plot', inasmuch as drama can be aligned with succinct summarisations of theme, might be more about diversity of form, topic, and content, which all underscore themes that are stitched together by improvisational performance. Regardless, it remains one of the representative and earliest examples of a 'petrodrama', which Macdonald defines as 'structured and conditioned by the oily machinations and social relations fuelling the extractive politics of the world's most inflammatory resource' (2015, 1).

In addition to the North Sea oil boom, showing logical historical and spatial sequences to contemporary events at the time of the early 1970s, *The Cheviot* was a timely performance within contemporary society. Britain in the 1960s witnessed rollbacks of progressive reforms instituted during the post-war period. Deindustrialisation, along with increased privatisation of social programmes aimed to support working-class groups who were most affected by consolidated industries, led to widespread unemployment and social stagnation. The gains made for working classes after the war period – through common publicly owned social institutions, transport, and property – were consistently being undermined by state-backed privatisation. This, against the backdrop of the Vietnam War, the Cold War, and the 1968 Paris resistance, created a contentious period in contemporary British politics that would eventually result in an apogee of conservatism in Thatcher's governments of the 1980s. However, British society and culture in the 1960s resisted many of these social stagnations by legitimising otherwise marginalised voices in society, particularly inspired by North American activist movements, which challenged unjust issues relating to gender, race, and class, and the ownership of land.

The 1970s in Britain experienced a range of problems: politically destabilising miners' strikes in 1972–1973, an IRA campaign, increasing unemployment, racial rioting at the Notting Hill Carnival, and a violent clash between police and picketers at the Grunwick film-processing plant in 1977. By the mid 1970s, Britain was no longer considered to be a 'world leader'; social stability rapidly abated and any sense of equitable affluence appeared to disappear and move to the top 7 per cent (Kershaw 2002, 133). Conservatism took firm root with Thatcher in 1979 and continued in the 1980s, disenfranchising the majority of people, part of the other 93 per cent. The once celebrated post-war prosperity and social tolerance no longer existed in Britain, leaving a vacuum for either progressive expansion or conservative austerity, with the latter winning this contest and pushing for neoliberal economic agendas that continue to have a hold in the present.

Within this socio-economic and political fabric, the 7:84 agitprop play *The Cheviot* continues to stand out as a daring and revelatory drama that encapsulates major social injustices over the last few hundred years, involving dispossession, increased capital for the few (i.e. the 93 per cent who own 16 per cent of the wealth when the play was first performed), and increased power for owners of capital, particularly multinational corporations who owned and controlled North Sea oil. *The Cheviot* influenced many other cultural responses to North Sea oil in literature and film culture, such as George Mackay Brown's novel *Greenvoe* (1973), Robert Alan Jamieson's novel *Thin Wealth* (1986), Bill Forsyth's film *Local Hero* (1983), Al Alvarez's book *Offshore* (1986), Jonathan Wills and Karen Warner's non-fiction book *Innocent Passage: The Wreck of the Tanker Braer* (1994), Lars von Trier's *Breaking the Waves* (1996), Ian Rankin's crime novel *Black and Blue* (1997), and, in part, Tom Morton's *The Further North You Go* (2003). This extensive list underscores only some of the creative works responding to a cultural self-definition of North Sea oil as it relates to Scotland.

McGrath's play attempts to tell Scotland's modern story and history of dispossession. *The Cheviot* addresses the issue of oil through agitprop, which is an artistic form challenging injustices through improvisational, touring, and interactive political theatre. The entire play confronts forms of spatial injustice, from land dispossession to speculative capitalism, and finally to a multinational petrostate, which disenfranchised the people of Scotland for decades, as well as Britain as a whole. Sticking with the theme of oil, however, I outline that the space-time compression of the 'laws of capitalism', or what the play also describes as the 'secrets of high industry', underlines the overarching theme (McGrath 2015, 98). In this socio-economic matrix of spatial and social injustices, I specifically examine the play as a petrodrama, exhibiting the political speed of the fossil fuel economy staged within an improvisational theatre performance, which demonstrates how the acceleration of economic processes in spaces of global neoliberal capitalism affect social lives and surrounding ecologies of place. The politics of spatial injustice directly relate to social and cultural production, which unfolds through *The Cheviot*'s theatrical form.

The 'laws of capitalism'

The Old Man functions in the play as a prophetic seer often used in Greek drama. He explains the 'laws of capitalism' for the audience in a longer monologue. He admits that before the Highland Clearances in the early nineteenth century, 'change had to come'. The population was growing faster than the limited methods of subsistence farming could support. Some people emigrated, but many remained. At the same time, the Industrial Revolution to the south in England continued to expand and accumulate wealth. As the Old Man goes on to describe, 'accumulated wealth had to be used, to make more profit – because this is the law of capitalism' (98). Accumulated wealth spread all over the world, particularly in British colonies, but Scotland remained an ideal space to implant technical innovation. The Cheviot sheep was subsequently introduced as a breed that could survive Highland winters and produce fine wool. The only problem was that the people had to first be removed to introduce the Cheviot; 'the law of capitalism had to be obeyed' (99). The technical innovations of the Cheviot contained more surplus value than humans living on the land.

Capitalism relies on simultaneously shrinking space and accumulative speed, what has often been referred to in economic theory as 'the political economy of speed', or the acceleration of politics and economies to gain power and control. The cultural theorist Paul Virilio has documented transformations of space and particularly in relation to speed within a post-Fordian and postmodern capitalist world-system throughout his prolific career. He argues that the notion of 'dromology', which is the study of the logic of speed, insists that velocity remains a core political and economic element in contemporary society and one that ultimately defines space (Virilio 1986). Virilio explains:

> To me, this means that speed and riches are totally linked concepts. And that the history of the world is not only about the political economy of riches, that is, wealth, money, capital, but also about the political economy of speed. If time is money, as they say, then speed is power. (cited in Armitage 2000, 35)

Speed is inherently the compression of time and space because it erases boundaries and constraints in human and geographical systems, such as regulation and protective oversight. Speed destabilises free movement of bodies (human and nonhuman) by industrialising them as part of a network of power through the control of space-time. Those who obtain wealth and therefore power expand around the globe and dictate the speed at which the world moves. This particularly holds true in specific cultures, some of which move at various speeds based upon their economic and social structures. The economic system of capitalism, and especially the velocity of neoliberalism, privileges those cultures and sociopolitical systems that continue to define space through speed as a society.

'Speed is not simply a matter of time', argues Virilio in an interview with John Armitage, but speed 'is also space-time. It is an environment that is defined in equal measure by space and by time' (Armitage 2001, 61). Virilio theorises speed as an environment, a setting and place, which measures time passing through movement between places. Movements in space define the domain of speed; movements can entail physical objects such as land or conceptual ones such as communication. For instance, the speed of oil moves at a much different velocity than art, with the former used as a controlling apparatus (both ideological and institutional) that governs space through velocity, while the latter reduces speed through human rather than technocratic perception as a conceptual process. One of Virilio's aims is to highlight the loss of 'human space' by reintroducing more existential and ecological spaces to experience 'human time' (Conley 2012, 83–4). This affect could also be defined as place-home, as a counter to the detrimental effects of solastalgia, where reduced economic speed increases human connection.

Adding to Virilio's 'political economy of speed', Michael Watts examines 'violence' as a result of 'fast capitalism' (2008, 8; Agger 1989). Fast capitalism underscores the accelerating movement of communication and commerce in an increasing information society. Such an extreme shift is why Rob Nixon's concept of 'slow violence' resonates as a way to explain environmental spatial injustices. The metaphor of moving slowly also functions literally and implies a spatio-temporal shift. Reducing fossil fuel production consumption must also be virtually eliminated in the cultural imagination, a move that would decelerate the speed at which the neoliberal economy moves, as well as force redefinitions of how the world-system might be restructured more equitably.

Within an economic system based upon speed, space and time are compressed to overcome the physical and market conditions of communication and transport. Marx initially claimed, 'Capital by its nature drives beyond every spatial barrier'. Creating 'physical conditions of exchange', according to Marx, whether it is through transportation or forms of communication, annihilates 'space by time' and thereby 'becomes an extraordinary necessity for it' (1973, 524). In other words, speed and mobility are linked to wealth through space, and this is particularly the case in the capitalist world-system fuelled by oil.

The period known as the Great Acceleration of the Anthropocene parallels the exponential growth of oil and gas production since the late 1940s, as well as surplus growth of capital, and it suggests in its name the speed at which the anthropogenic environmental crisis has continued unabated. The current capitalism world-system largely funded by oil remains a spatial issue, while at the same time the ways in which the longer history of petrocapitalism exploits humans and environments remains a social justice issue. These two threads underlie the tragicomedy and dramatic tension in *The Cheviot*, a play McGrath hoped would displace the 'dominant bourgeois culture of late-capitalism', or what we might also call the neoliberal culture extending from petrocapitalism, by staging before audiences the immense speed of this process.

The logic of speed within a history of petrocapitalism (and petromodernity) also relies on the anxiety and opportunity dialectic fundamental to oil in order to promote forms of velocity. The character Texas Jim in the oil section of the play exemplifies this dynamic. The instability of acceleration, particularly of socially produced formations that hold the welfare of people in the balance, creates feelings of excitement leading to speculation (opportunity), while also influencing decisions based upon fear rather than rationality (anxiety). This spatial tension between time and space underlines the history of petromodernity, where capital produces speed that then ignites power and control of resource-based economics (largely through fossil fuels). The speed of global events in an ever-increasing scale nullifies the social and cultural in both meaning and value.

Using the limited space of the theatrical stage as a mode of contestation, *The Cheviot* confronts 'the secrets of high industry' by using capitalism's own strategy: to compress time and space within a play, linking 250 years of history (since the 1745 Jacobite Rebellion) and various genres and modes of storytelling in a democratised form of social theatre for the people. The chronology of the play appears linear, but the underlining elements of capitalism that replicate archetypal systems as process – repeating and recycling – ultimately spans across time and space (Macdonald 2015, 56). The historical pacing of the play forces audiences to exist simultaneously in multiple histories, which has the effect of holding the Highland Clearances in juxtaposition against the more contemporary Oil Clearances.

In this way, both the form and the theme of *The Cheviot* accentuate the spatial injustices of oil in Scotland. This dynamic is evident throughout the play, but it first appears at the beginning with the historical figures James Loch and Patrick Sellar and their economic strategy known as the Highland Clearances, beginning in 1813 and escalating to forced evictions by 1814. The Clearances were supported by the Dukes of Sutherland and based upon the realisation that cultivating Cheviot sheep on the land generated more profit than people, an economic insight that began the forced dispossession of the Highlanders from their homes.

In the first section of the play, Loch, who is with Sellar as an advocate for the Sutherland Estate and primary architect of the Clearances, argues, 'To be happy, the people must be productive' (McGrath 2015, 90). While productivity equates to speed, it is also an argument to increase capital. Non-accelerated time (i.e. 'slowness'), which is a contrasted cultural mode of traditional Highland culture, is what Loch refers to as 'a form of slavery' of 'their own indolence'. He proclaims, 'Nothing could be more at variance with the general interest of society and the individual happiness of the people themselves, than the present state of Highland manners and customs' (90). Stellar chimes in approvingly to declare the people of the Highlands should move to the coast and work for industry. Creating a draconian zero-sum situation, he maintains they must 'worship industry or starve'. Their present 'enchantment' of living in the Highlands, according to Sellar, 'must be broken' (90).

The velocity at which Loch and Sellar want to move across the spaces of the Highlands, subsequently forcing the Highlanders to move as well, links to the profit they will receive once the clearances are complete. Sellar boasts that the

> highlands of Scotland may sell £200,000 worth of lean cattle this year. The same ground, under the Cheviot, may produce as much as £900,000 worth of fine wool. The effects of such arrangements in advancing this estate in wealth, civilisation, comfort, industry, virtue and happiness are palpable. (90)

This entire discussion eventually breaks into song as a way to visually showcase their argument. The satirical song is a capitalist-infused duet – drawing on the original version of James (Jimmy) Copeland's 'These Are My Mountains' – that recurs in various iterations throughout the play:

> As the rain on the hillside comes in from the sea
> All the blessing of life fall in shower from me
> So if you'd abandon your old misery –
> I will teach you the secrets of high industry:
> Your barbarous customs, though they may be old
> To civilised people hold horrors untold –
> What value a culture that cannot be sold?
> The price of a culture is counted in gold. (91–2)

Although related to the Highland Clearances, this scene demonstrates the difference in speed at which industry and culture moves. There are obvious spatially unjust scenarios related to the Highlanders' displacement, where the police and military would forcibly and often violently removed people from their homes (including woman and children) and then burn them to assure people do not return. Even Sellar's name, despite it being historically accurate (i.e. Patrick Sellar), serves to pun his own actions. He is a 'seller' of Highland culture and land; anything can be bought and sold in capitalist 'industry' where the 'price of a culture is counted in gold'. As Marx earlier observed, 'First the labourers are driven from the land, and then the sheep arrive' (1976, 556).

This scene between the two industrial capitalists in the nineteenth century foreshadows later encounters with the resource/corporate capitalist Texas Jim, who openly concedes he will collude with the British government (i.e. Whitehall) to steal oil from the locals and make enormous profits for his company and himself. Although more of a symbol than a person, Whitehall serves as a metonym for the governmental buildings and departments in central London. The detached name assumes a spatial proximity between London and the Highlands, while it is also juxtaposed against Texas Jim – an absurd figure, serving as a caricature or PR agent on behalf of multinational petroleum corporations. Both figures mirror a recurring history as represented by Loch and Sellar, and later, in part, through Andy and Lord Vat in discussions about selling off the Highlands for tourism to hunt 'stag' in the second section of the play.

Whitehall and Texas Jim both sing another skewed variation of 'These Are My Mountains', but now retitled 'No Land's Ever Claimed Me':

> As the rain on the hillside comes in the from the sea
> All the blessings of life fall in showers from me
> So if you'd abandon your old misery
> Then you'll open your doors to the oil industry – (151)

At this point, the 'Strathnaver Girls', appearing on stage as accompanying vocals commonly used in 1960s pop music, twice sing the chorus: 'Conoco, Amoco,

Shell-Esso, Texaco, British Petroleum, yum, yum, yum' (151). These five oil companies not only controlled global petroleum production in the early 1970s, but they also, as the play indicates, manipulated governments and economic activity (represented by Whitehall). The pop rock theme between Texas Jim and Whitehall parallels the edgy and sexy aspect of oil and the prospects it might bring to Scottish culture, which is seen as both dangerous and opportunistic.

There is tacit assumption by the character Texas Jim that what he sings might not be completely understood by the audience. He seizes this opportunity like a huckster. Moving back to the song with Whitehall, Texas Jim continues to sing alone:

> There's many a barrel of oil in the sea
> All waiting for drilling and piping to me
> I'll refine it in Texas, you'll get it, you'll see
> At four times the price that you sold it to me. (151)

The Girls sing the chorus again and then only Whitehall assures the Scottish people that jobs will appear because with the 'building of oil rigs and houses' there will be 'a boom-time a-coming'. Texas Jim then pours symbolic drinks of oil while Whitehall sings, 'let's celebrate – cheers –'. Texas Jim concludes the song in a more sinister tone, and somewhat contradictory to what Whitehall just explained to the Scottish people, by confirming 'the Highlands will be my lands in three or four years' (152).

After this musical interlude between Texas Jim and Whitehall, the Aberdonian Rigger (A.R.) offers a sober moment of clarity in contrast to the farcical song and dance. In fact, the A.R.'s speech resembles a documentary-style interview focused on one person discussing his personal experience.[2] The speed of the play at this point shifts from fast-paced to silent and slow. In direct contradiction to what the audience just heard from Whitehall, the A.R. discusses the reality of working on the rigs, where riggers work 84 hours a week in 12-hour shifts (two weeks on and one week off); the oil companies do not pay overtime or sick pay and they can sack a worker without cause. In addition, the petroleum companies who monopolise the North Sea will not take 'union men' with labour rights. The political economy of speed from the North Sea oil boom resulted in loss of life through helicopter and driving accidents, damage to enhanced speed of existing communities in the Highlands, and degradation to local and regional ecologies (Harvie 1994, 261).

The A.R. explains that at the 'first sniff of oil, there was a crowd of sharp operators jumping all over the place buying the land cheap. Now they're selling it at a hell of a profit' (153). As the song and dance disappear, moving from the imagined to the real, the A.R. discusses the unsustainable speed at which the workers must move. The function of the solitary voice without accompanying music or dance stresses the extreme absurdity of the accelerating effect of the oil industry (embodied in the equally ridiculous song with Texas Jim), and the subsequent impacts of speed on the people and place of the Highlands.

The oil companies refuse to negotiate and seek compromise to obtain the land from famers in Aberdeenshire. Rather, as Texas Jim boasts, they have already purchased them through the willingness of the government. One of the theatre's company/community members, citing Willie Hamilton, MP, states that there 'is a great danger of the local people being outwitted and out-manoeuvred by the Mafia from Edinburgh and Texas' (153). Other company members on stage chime in resonant lines that have related to the theme of the entire play: 'The people must own the land. The people must control the land. They must control what goes on it, and what gets taken out of it' (153).

Who owns and maintains control of the north of Scotland adjoining the North Sea? *The Cheviot* revisits this recurring question throughout the play, demonstrating that first the interests of empire and then the shareholders of multinational corporations have systemically exploited the land and people. As suggested at the beginning of this section, *The Cheviot* and all of 7:84 productions challenge ideas of place, capital, and power through a democratised form of community theatre. The historical loss of place-home through dispossession endures to the present moment for Highlanders, with the accelerating and exploitive effects of capital through economic and political power. The company members, who serve as voices of the community, speak truth during the outbursts of pantomime, song, and dance. The community members ask the question of who owns the land: 'Marathon Oil? Apco of Oklahoma? Chicago Bride and Iron of Chicago? Cleveland Bridge and Engineering?' (154). Another company member comments, 'Scottish capitalists are showing themselves to be, in the best tradition of Loch and Sellar – ruthless exploiters'. The same person adds, 'Nationalism is not enough. The enemy of the Scottish people is Scottish capital, as much as the foreign exploiter' (155).

Lord Polwarth, who was Minister of the State in 1972 (which includes management of oil), but who was also the former Governor of the Bank of Scotland and Chemical Industries Director, attempts to assuage the community's discontent by dryly confirming that the people of Scotland 'will benefit from the destruction of their country' (155). Once he sings a song about oil, Texas Jim and Whitehall satirically turn Polwarth into a puppet by holding imaginary strings from his wrists and back. They all sing an altered version of 'Lord of the Dance': 'Oil, oil, underneath the sea, / I am the Lord of the Oil said he, / And my friends in the Banks and the trusts all agree' (156).

Speaking on behalf of Texas Jim and Whitehall, Polwarth acknowledges, 'I came up from London with amazing speed / To save the Scottish Tories in their hour of need . . . Now I am a man of high integrity / Renowned for my complete impartiality' (156). After the song, Texas Jim and Whitehall release the invisible puppet strings and Polwarth falls limply to the ground. The dizzying revolving door among multinational corporations (Texas Jim), the British government (Whitehall), and the National Bank of Scotland (Polwarth) underscores not only the collusion between power and capital, but also the desire to move rapidly with 'amazing speed', mirrored by the pace and hurried action of the play and accompanying music.

Texas Jim clearly signifies corporate oil. He is a contemporary charlatan selling a product of possibility and threat to what are considered to him a gullible audience/country because they cannot live by nor understand capital's incessant speed. In his opening song, which is a fiddle-based hoedown for square dancing, Texas Jim hastens his voice along with the tempo and, by the final crescendo, hysterically reveals the truth as he spins out of control:

> You play dumb and I'll play dirty
>
> All you folks are off your head
> I'm getting rich from your sea bed
>
> I'll go home when I see fit
> All I'll leave is a heap of shit
>
> You poor dumb fools I'm rooking you
> You'll find out in a year or two. (148)

McGrath hyperbolises Texas Jim to magnify the otherwise hidden agenda of petrochemical companies, which is to manipulate local people out of their land, particularly if they remain in the way of potential oil profit. Another company member in the play confesses, 'As in all Third World countries exploited by American business, the raw material will be processed under the control of American capital – and sold back to us at three or four times the price –' (150). Again, the underlying current of this business practice that exploits workers and local people around the world is the ungovernable movement of speed within socially produced spaces. For instance, Texas Jim speaks quickly, moves abruptly, and sings songs that contain fast tempos, all of which parallel the speed of the oil industry he symbolises. Macdonald comments that Texas Jim's 'suddenness, speed and accelerated action' are used to transform 'the environment' in their favour, what could be the environment of the North Sea or on stage, which is obvious in Jim's desperation about making this deal while working through the musical interlude (2015, 23). As a representative corporate voice, Texas Jim agrees, 'Yes sir, and we certainly move fast' (149).

Speed displaces people and creates unsustainable social conditions; it also counters ecological stability and eliminates any sense of place-home for the people who desire localised relationships to the land. The irony here is the Highlanders in *The Cheviot* live a form of eco-localisation, which is an economic system and way of being in the world that would counter the time-space compression of neoliberalism and the cause of climate breakdown, polluted air, lack of food, diminishing drinking water, and so on (North 2010, 585). The Highland Clearances replaced subsistence farming, which was admittedly providing a limited amount of food for the populations in the early nineteenth century, with Cheviot grazing to produce more capital for England. Grazing had adverse effects upon the ecologies of the Highlands. But oil, as discussed in the previous chapter, continues to produce the same displacement and oppression as before, but now with the compounding problem of rising global temperatures.

Because the play was written and performed in the early 1970s, there is no direct mention of changing climates caused by increasing amounts of greenhouse gases into the atmosphere. However, there is foresight toward this outcome in the scene between the Crofter and his Wife, who both own a bed and breakfast to accommodate the petrotourism in the north of Scotland amid the oil boom of the 1970s. The Crofter comments that their guests from Rotherham will 'be looking like snowmen stuck out there in this blizzard –'. The Wife responds, 'Och, it's terrible weather for July', at which point the Crofter acknowledges, 'It's not been the same since they struck oil in Loch Duich' (159). The cause and effect of drilling and climate might be hard to prove at the time of the play, but it nevertheless anticipates unusual weather patterns resulting from carbon and other pollutants in the atmosphere.

The Crofter and his Wife experience solastalgia because they feel homeless or exiled without ever leaving the Highlands. The Crofter, who was injured working on the oil rigs, resigns himself to the chair of his house to watch the oil rigs each day on the horizon of the North Sea. This scene showcases the immobility of those who have been displaced during the oil boom. The demand for property guided by inflated oil capital for mortgages and rents dispossessed many local communities to other parts of Scotland. The Crofter and his Wife have little financial flexibility to keep their home, so they open a bed and breakfast ironically for petrotourism, which supports workers in the industry and people visiting the 'spectacular' oil rigs (158). The irony is obvious: petrotourism is a business that supports the industry that injured and then financially and socially displaced the Crofter and his Wife.

Throughout their dealings with the bed and breakfast, the Crofter and the Wife continually observe and even experience (through the changing climate) ecological effects that create distress. As the Wife appeals to the tourists, she truthfully admits that 'on a clear day' you will be able 'to see the oil-rigs – oh, they're a grand sight, right enough'. She confesses, albeit in a tragicomic tone, 'Aye, you'll get a much better view now the excavators digging for the minerals have cleared away two and a half of the Five Sisters of Kintail' (159). The Five Sisters are a pristine mountain range on the north-west Highlands of Scotland owned by the National Trust for Scotland (NTS). The conflict is clear throughout this scene. The Crofter and Wife depend upon the tourist incomes because of the inflated property prices triggered by an accelerated oil economy in the Highlands. But, at the same time, they understand the consequences of such an industry. They acknowledge the ecological changes with the blizzard in summer and the environmental destruction of Kintail.

In addition, *The Cheviot* confronts the lack of social sustainability in a system of fossil capitalism (and increasing forms of neoliberalism) and demonstrates how this process impacts the loss of place. The Clearances both symbolically and literally made way for oil exploitation, and in this way the play functions as one of the earliest forms of a petrodrama, staging as a performative and literary form the spatial injustices that oil produce around the globe. But there are also some of the clearest confrontations of ecological concerns wrapped up in 'the black, black oil' section at the end of the play. In Texas Jim's opening hoedown, he basically outlines the ecological impacts of this operation. He sings, 'So leave your fishing, and leave your soil / Come work for me, I want your oil'. This results in what will happen: 'Screw your landscape, screw your bays / I'll screw you in a hundred ways –' (147).

What is being sold to the Scottish people differs from the ecological and social outcome – that is, fossil fuels increase carbon in the atmosphere and pollute water and soil, affecting local communities more directly than those profiting off of them. Texas Jim reveals this in the song during a moment of hubris and excitement when he concedes how he, as a symbol of oil companies, intends to 'screw' the people. This should come as no surprise to contemporary audiences of the play in the 1970s, but more particularly it resonates with audiences who attended the recent reproduction of *The Cheviot* put on by the Dundee Rep Theatre in 2015.[3] Petrochemical companies have historically funded billion-dollar campaigns to reinforce scepticism about any climate- or pollution-related dangers resulting from extraction, production, or transport of oil and gas. Concrete evidence now exists that petrochemical companies have known about and admittedly suppressed public knowledge about the dangers of climate change since at least 1977. This proof has become transparent through the United States court case against the world's largest oil and gas company, ExxonMobil (Goldenberg 2015; Hall 2015).

A story without an ending

The Cheviot elucidates environmental issues indirectly through a socio-economic lens, one that magnifies social change in smaller communities as a result of the accelerated capitalist world-system. The play challenges social constructions of space through three distinct phases of land control of the Highlands; it also re-establishes a bond between place and culture through the humanities. By using social theatre as a way to confront people's conceptual systems, McGrath reframes energy narratives as a social and cultural question. To this effect, people can change their values and understanding about energy and environmental dilemmas, which directly relates to habits, institutions, beliefs, and power. Who owns the land? Who owns the oil? Who controls and maintains the natural, humanistic, and cultural resources of Scotland? Awareness of our conceptual systems make the insidiousness of Texas Jim all the more apparent, as McGrath emphasises in the play. The renewable energy of social theatre confronts the limited energy of fossil fuels both in form and social outcome. *The Cheviot* reclaims the power of oil through the social order, one that has continued into the present in part because of creative production in the humanities.

Macdonald argues that literature

> in its various modes, genres, and histories, offers a significant (and relatively untapped) repository for the energy-aware scholar to demonstrate how, through successive epochs, particularly embedded kinds of energy create a predominant (and oftentimes alternative) culture of being and imagining in the world. (2013, 4)

The Cheviot serves as a relevant example of contemporary Scottish literature that surveys 'successive epochs' in various histories and genres. McGrath's play continues to resonate to this day. The 2014 Scottish Independence Referendum to leave the UK serves as a political gesture that is likely to once again appear in response to the 2016 United Kingdom European Union Referendum (i.e. Brexit). While yearning for Scottish independence speaks to a wide range of concerns, Scotland's 2014 Referendum and ongoing desire to leave the UK is largely about economic independence and future control over the North Sea, even if the future of North Sea energy production moves to renewables (such as tidal power off of Orkney) in the wake of dwindling oil and gas revenues. It is even estimated that Scotland contains the potential to produce 25 per cent of Europe's tidal and wind power. The European Marine Energy Centre (EMEC) has constructed the world's most powerful tidal turbine off of Orkney. It aims to generate enough electricity to power around 175,000 homes. The future control of this energy source remains controversial in the shadow of North Sea oil's history for Scottish autonomy.

Following the sustainable model of Norway, Scotland would ideally have state ownership over their energy resources and interests in the wake of a long history of disenfranchisement and displacement. In contrast to the UK, Canada, the US, Russia, or any other oil-producing nation, Norway wisely socialised oil and gas production. Because of this economic foresight, every Norwegian has become a millionaire (in kroner), which equates to £100,000 (Chakrabortty 2014). In addition, oil and gas revenues have been used to invest in alternative forms of energy for future generations, with the hope that they will be a post-carbon country by 2030. *The Cheviot* reminds audiences to consider where would Scotland be now if they had jurisdiction over North Sea oil over for the last 50 years?

If all of the North Sea oil and gas revenues were invested in safe assets, according to John Hawksworth, a chief economist at WaterhouseCoopers, then profits for the UK would have been worth £450 billion by 2008 (and this is a conservative estimate). Broken down, this amounts to £13,000 per person in the UK, not to mention what this would be worth for every person in Scotland had they complete control of the North Sea oil and gas revenues over the past 40 years (Chakrabortty 2014). Hawksworth sarcastically titled his essay on the subject, 'Dude, Where's My Oil Money', to assert the inequitable collusion of capital and power that produces wealth only for the 7 per cent, not the majority 93 per cent (to use McGrath's figures in the late 1960s, but which are much worse now: 1:99). The tax cuts for the public sector, which did not avail as much from oil profit, increased public deficit, thereby promoting unjust austerity measures for hospitals, public transport, and social assistance programmes. As Hawksworth asserts, 'the oil money enables non-oil taxes to be kept lower', which supports the political optics of tax cuts (Chakrabortty 2014). In this sense, the UK experienced an unprecedented economic increase for every citizen throughout the last 40 years, but all of this capital was diverted to wealthy and private interests, many of which are not part of the UK.

As the play claims at the onset, 'we'll be telling a story. It's a story that has a beginning, a middle, but, as yet, no end –' (85). There is no ending because until spatial injustices can be resisted and transformed, the social dynamics staged in the play and the speed at which they move will continue. *The Cheviot* galvanises communities to mobilise against injustices that have occurred for many centuries, remaining in the contemporary moment while also reverting to the past. Neoliberalism siphons powers of the state to divert more resources to the wealthy; this is the fundamental ideology of privatisation. Compressing time and space through the political economy of speed enables this perpetual cycle, resulting in social injustices associated with space and place in *The Cheviot*, such as contested displacement (clearings), economic disenfranchisement (losing homes or raising home prices and/or rents), and loss of place-home (through environmental degradation). All of these examples result in various forms of ecological exile. Mismanagement of North Sea oil signals the moribund and inequitable economic practice of neoliberalism, which produces billions of tax cuts for the wealthy, but leaves the state reduced to emaciated austerity budgets. This is the legacy of North Sea oil in Britain and especially in Scotland.

The Cheviot ends by reminding audiences it could be different: 'Now the Black Black Oil is coming. And must come. It could benefit everybody. But if it is developed in the capitalist way, only the multi-national corporations and local speculators will benefit' (162). But, as Monbiot claims in the previous chapter, will the collective dying corpses of the intricately linked neoliberal and fossil fuel economies that have 'failed us' finally die or continue on like zombies (2016)? *The Cheviot* directly challenges this 'anonymity' that Monbiot argues has allowed neoliberal economics to survive, as well as the petrochemical revenues that fund them (2016). Slowing down the speed at which society functions by focusing on non-technocratic human places highlights the spatial injustices related to ecological and energy themes within the humanities. *The Cheviot* demonstrates both the speed and slowness of petrospaces through its theatrical form and socially reconstructed response to North Sea oil and gas from the Scottish people.

Notes

1 As Harvie points out, the Scottish cultural renaissance of the 1970s and particularly the 1980s parallels the Irish Cultural Revival beginning in 1891 with the Gaelic League, and eventually spawning the Gaelic Athletic Association (GAA), Abbey Theatre, Sinn Féin, and other literary and cultural developments ultimately contributing to Irish independence in 1921. Harvie's comment pinpoints that culture in the form of artistic production historically responds to and alters political and economic injustice, which for Scotland persisted for centuries (as *The Cheviot* outlines), culminating in contemporary North Sea oil and gas production (1994, 277).

2 A film version of *The Cheviot* was made in 1973 with the same cast as the touring play. During the film, it showed interviews with actual riggers instead of cast members in the play speaking these lines (Mackenzie 1974).

3 While attending a rare performance of *The Cheviot* by the Dundee Rep Theatre in September 2015, I noticed that the audience became much more participatory during the oil section of the play. This performance also occurred a year after the results of the 2014 Scottish Independence Referendum, and the audience in at the play in Dundee appeared to be pro-independence 'Yes'.

Bibliography

Agger, Ben. 1989. *Faster Capitalism: A Critical Theory of Significance*. Urbana-Champaign, IL: University of Illinois Press.

Armitage, John, ed. 2000. *Paul Virilio: From Modernism to Hypermodernism and Beyond*. London: Sage.

——, ed. 2001. *Virilio Live: Selected Interviews*. London: Sage.

Buiter, Willem H., and Marcus Miller. 1981. 'The Thatcher Experiment: The First Two Years'. *Brookings Papers on Economic Activity* 2: 315–79.

Chakrabortty, Aditya. 2014. 'Dude, Where's My North Sea Oil Money?' *The Guardian*, 13 January. Accessed 3 November 2016. www.theguardian.com/commentisfree/2014/jan/13/north-sea-oil-money-uk-norwegians-fund.

The Cheviot, the Stag and the Black, Black Oil. 1974. Directed by John Mackenzie. London: BBC.

Conley, Verena Andermatt. 2012. *Spatial Ecologies: Urban Sites, State and World-Space in French Cultural Theory*. Liverpool: University of Liverpool Press.

Goldenberg, Suzanne. 2015. 'Exxon Knew of Climate Change in 1981, Email Says – but It Funded Deniers for 27 More Years'. *The Guardian*, 8 July. Accessed 10 November 2016. www.theguardian.com/environment/2015/jul/08/exxon-climate-change-1981-climate-denier-funding.

Hall, Shannon. 2015. 'Exxon Knew about Climate Change Almost 40 Years Ago'. *Scientific American*, 26 October. Accessed 10 November 2016. www.scientificamerican.com/article/exxon-knew-about-climate-change-almost-40-years-ago.

Harvie, Christopher. 1994. *Fool's Gold: The Story of North Sea Oil*. London: Penguin.

Hobsbawm, Eric. 1999 [1968]. *Industry and Empire: From 1750 to the Present Day*. London: Penguin.

Kershaw, Baz. 2002 [1992]. *The Politics of Performance: Radical Theatre as Cultural Intervention*. London: Routledge.

Marx, Karl. 1973. *Grundrisse: Foundations of the Critique of Political Economy*. London: Penguin.

——. 1976. *Capital*, vol. I. New York: Vintage.

Macdonald, Graeme. 2013. 'The Resources of Fiction'. *Reviews in Cultural Theory* 4(2): 1–24.

——. 2015. 'Commentary'. In *The Cheviot, the Stage, and the Black, Black Oil* by John McGrath, edited by Graeme Macdonald, 17–80. London: Bloomsbury.

McGrath, John. 1980. 'The Theory and Practice of Political Theatre'. *Theatre Quarterly* 9(36): 43–54.

——. 2015 [1981]. *The Cheviot, the Stag, and the Black, Black Oil*, edited by Graeme Macdonald. London: Bloomsbury.

Monbiot, George. 2016. 'Neoliberalism: The Ideology at the Root of All Our Problems'. *The Guardian*, 15 April. Accessed 25 October 2016. www.theguardian.com/books/2016/apr/15/neoliberalism-ideology-problem-george-monbiot.

North, Peter. 2010. 'Eco-Localisation as a Progressive Response to Peak Oil and Climate Change: A Sympathetic Critique'. *Geoforum* 41: 585–94.

Virilio, Paul. 1986 [1977]. *Speed & Politics: An Essay on Dromology*. New York: Semiotext.

Watts, Michael. 2008. *Struggles over Geography: Violence, Freedom, and Development at the Millennium*. Hettner Lectures no. 3, University of Heidelberg, Department of Geography.

5 Sullom Voe

Place and memory in petrospaces

The previous chapter focused on North Sea oil and Scottish Highland culture, but this chapter moves further north to examine the relationship between oil and memory through literature and media culture in the Shetland Islands (also referred to as the Shetland Isles or Shetland). Outlying as it may seem, Shetland has existed in the North Sea oil and gas industry since the 1970s. This subarctic archipelago, blending Scottish and Nordic history and culture but part of the United Kingdom, has resisted various phases of colonisation over the last several hundred years. Although slight in population, Shetland contains a productive and responsive literary history to oppression and occupation. Over the last 40 years, writers and poets have specifically responded to the oil and gas industry, showing both the benefits and difficulties of living in the centre of the North Sea oil boom. This chapter aims to link place-based poetry, film, and Web-based media in Shetland's petroculture through a discussion about Roseanne Watt's filmpoem *Sullom* (2014).

Sullom is part of a larger online and open-access collective entitled *Documenting Britain*. Started in 2013, *Documenting Britain* seeks to trace people, environments, and culture in the 'British Isles' through various creative works of art, film, music, and literature. The website functions as a critical and creative record of the archipelago of the British Isles. *Documenting Britain* serves as a constellation of Web-based literature and media, which represents an emerging platform to publish literary and visual works.[1] By publishing literature or visual culture on an open-access website, rather than traditional print or digital formats, writers, filmmakers, and poets seize the immediacy of social action through their art. The current democratised forms of open access media allow for activist-oriented artistic forms to address critical environmental issues. Using online platforms to disseminate artistic production is a fundamental aspect of the environmental humanities in the twenty-first century.

Watt's contribution to this extensive online artist collective uses poetry of place in the Shetland Isles. However, some of her work also combines visual images and short films, the latter of which Watt calls 'film portraiture' because it documents autobiographical links to place and memory. Besides *Sullom*, Watt's portion of the *Documenting Britain* website contains two other poetic-prose works entitled

'Mell-Moorie' and 'Yurden – A Circle Poem', both of which record Watt's film portraiture to place. The filmpoem *Sullom* exhibits a compelling approach in contemporary culture to map part of the history of oil and gas exploration and production in Shetland through the intersection of literature and media. In the following pages, I consider how *Sullom* unsettles dominant histories of displacement in the Shetland Isles by reframing the narrative through poetry about place, thereby confronting environmental and spatial injustices within the history of North Sea oil. To begin, the opening of this chapter briefly works through Shetland's cultural and environmental relationship to North Sea oil.

Shetland's place-based petrohistory

The name *Sullom* derives from an inlet called Sullom Voe, which is between the North Mainland and North Mavin. Sullom is a county in Shetland and Voe is the word for a small bay or narrow creek in the Shetland and Orkney Islands. Sullom Voe is also the name of an enormous oil terminal used for storage, refining, and trans-shipment in this inlet. Serving as a petrospatial coordinate, as much as place connected to local history and culture, Sullom Voe is both culturally and geographically interlinked with North Sea oil. As stated in in the previous chapter, oil was first discovered in the North Sea in 1969. The unearthing of oil prompted immediate construction of the Sullom Voe oil terminal. It was operational in 1972 and completed in 1982; at that time, it was the largest crude oil terminal in all of Europe. Sullom Voe continues to refine and store oil and liquefied gas from offshore sites in the North Sea. These fossil fuels are then transported around the planet by enormous tankers.

Sullom Voe has been both a benefit and detriment to Shetlanders, as well as to Britain as a whole, which has garnered responses from many Scottish and Shetlander writers and poets. Graeme Macdonald maintains that 'the control – and squandering – of North Sea oil remains one of the most controversial issues in modern British politics' (2015, 55). This contemporary petrohistory remains contentious for Shetlanders, a community of islanders that for centuries relied on fishing industries (catching, farming, and processing) and raising livestock. The shift occurred in the twentieth century when they were forced to embrace the North Sea oil boom (and bust) because of economic stasis caused by overfished waters largely due to large-scale global fishing industries. By as late as 1982, thousands of tons of crude oil were transferred by pipeline from the North Sea oil wells and stored in enormous tanks on Sullom Voe.

Lerwick, the capital and administrative centre of Shetland, initially became a hub of offshore service and supply to support nearby oil platforms. Shetland's main airport is called Sumburgh. The function of the airport quickly became a way to move people and materials to and from the oil platforms, although because the demand became so enormous, another airport called Scatsta was constructed near Sullom Voe as oil operations increased. Because of the quick pace in development, returning to the effect of acceleration of time-space compression (see Chapter 3), much of the land where smaller Shetland communities lived and worked was negatively affected throughout this process. Shetland topographies were, for instance, used as fabrication yards for massive steel and concrete production sites (House 2003, 14). Because of this constant activity on Shetland, with a population of 22,768 in 1981 (which only grew to 23,200 in 2013), it was difficult to separate distinct and diverse threads of Shetland culture or identity from the petroculture that dominated the islands for over five decades (*Shetland in Statistics* 2014).

To cite one cultural example, the BBC series *Shetland* (2013–) attempted to limit the oil and gas identity often attached to Shetland. The television series, which is a murder mystery show similar to that of *Wallander* (2008–2016) or *Inspector George Gently* (2007–) focused on crime in a specific place, only has one episode dealing with Shetland's relationship to oil and gas in the first three seasons. This may seem like a curious effort to resituate or even rebrand Shetland as a place for tourism rather than a destination known for North Sea oil and gas production (at least in the last 50 years). While Shetland's economy was forced diversify, relying more on tourism, textiles (knitwear), creative industries (arts and culture), and food, it continues to embrace oil and gas even with significant reductions in exploration and production. This effect is largely because of the politics of Westminster in London, which, along with multinational petrochemical companies, largely controls Shetland's oil and gas industry.

Despite passing the peak oil threshold, as well as falling revenues coupled with eliminating thousands of jobs, oil and gas construction in Shetland aims to expand. In 2015, the French oil and gas company Total opened a new gas plant at Sullom Voe, which connects to the massive Laggan-Tormore Pipeline network. This £500 million investment (despite the global downturn in oil prices) would convince some people that a future of oil and gas remains in Shetland. British Petroleum (BP) has reportedly proposed a redevelopment plan of the main oil terminal at Sullom Voe. The expected net benefit of the Laggan-Tormore Pipeline would be that it could supply power to all of Scotland, while also providing up to 800 jobs ('Total's Shetland' 2016). These are marketing propositions and projections, however. In January 2016, BP announced that they are shedding 600 jobs from the North Sea oil operations. At its peak in the early 1980s, the North Sea generated £12 billion in government revenues for Britain, compared with £4.7 billion today. It is estimated that about 23,000 jobs (more than the population of Shetland) will be eliminated in the North Sea over the next five years. These jobs are likely to never come back (Stevens 2016). With dwindling oil and gas prices globally, as well as a limited supply of oil and gas in the North Sea and with many other economically viable alternative energy options, investing in fossil energy futures seems irrational and ultimately futile.

Many of the proposed expansions of petrochemical industries fail to emphasise the environmental costs of oil production. The Earth is close to passing a 2°C increase in global warming. In September 2016, the world passed for the first time 400 parts per million (PPM) of carbon dioxide (CO_2) levels, which is the threshold scientists claim is necessary to remain under (if not 350 PPM) to avoid catastrophic climate change (see more in Chapter 7). Such an increase happened more rapidly than climate scientists predicted it would. Extracting and processing fossil fuels are primary factors for increased carbon in the atmosphere leading to global warming. Investing further in unsustainable and ecologically damaging forms of fossil fuel energies, especially when we have past the point of 'peak oil' and are living in the hottest period in recorded history, ignores scientific and public support for transitioning into renewable energy sources that would likely prevent catastrophic consequences.

Construction of such massive sites creates other environmental problems beyond extracting oil from the Earth and increasing carbon in the atmosphere. Located in north-west Shetland, the Laggan-Tormore oil and gas fields connect by pipeline to the Shetland Gas Plant, which is the largest construction project in the UK outside of the 2012 London Olympics ('Total's Shetland' 2016). The Shetland Gas Plant is built on a peat bog next to Sullom Voe, which resulted in the decimation and removal of enormous areas of peatland on the island.

While some might consider peatland wasted space, they are actually important ecosystems in preventing climate change. Peatland serves as natural carbon sequestration units, also called 'sinks', which are places where CO_2 is captured and 'sequestered' from the atmosphere. This process, known as carbon capture and storage (CCS), remains a temporary solution for reducing carbon in the atmosphere. Over 20 per cent of the world's current terrestrial carbon is captured and stored in peatlands of the northern hemisphere ('Climate Change' 2007). The destruction of such wetlands not only eliminates surface area where carbon is captured and sequestered; it also releases stored carbon back into the atmosphere (Gladwin 2016, 228–9). There are extreme environmental effects of building more refineries on Shetland, as well as pipelines to transport oil and gas, particularly when it involves removing enormous expanses of peatland.

The energy industry in Shetland is currently undergoing change. An alternative future of North Sea energy production is already moving to renewables, such as tidal power off of Orkney and wind energy on Shetland. Scotland has the potential to produce 25 per cent of Europe's tidal and wind power, and these tidal power generators are in the North Sea off of Orkney and Shetland.[2] Positioned on Bluemull Sound between the islands of Unst and Yell, the company Nova Innovation just activated the second of five 100 kW tidal turbines that would provide electricity to parts of Shetland and the UK. Despite increased efforts to supplant oil and gas with tidal and wind power in the North Sea, petrochemical companies such as Total fund expansion programmes to extend the reign of oil and gas production.

The question remains, however, how do writers and artists in the humanities depict the environments they know while also working to change them? How can they socially reconstruct or reclaim the petrospaces in and around Shetland? This chapter examines spatial injustices of oil and gas in Shetland and how they can be addressed and altered through forms of literary and media culture, as we see in Watt's filmpoem *Sullom*, by re-conceptualising perceptions of Sullom Voe through a combination of literary and visual narratives of place that reclaim a way of being in the world from the dominant petroculture in which they function.

Spatial bodies and holographic memory

In the filmpoem *Sullom*, Watt frames the spaces of Sullom Voe in two main ways: through the real and imagined body and the multidimensional effect of the visual and poetic hologram. This bi-medial relationship (as an image-text) connects the poem (as a literary text) to the film (as a media text). The filmpoem's musical score provides an additional element that creates an anti-aesthetic effect, ironising advertisement campaigns placed by energy companies. To begin, however, I will first provide a quick overview of the filmpoem. The following pages will discuss the bi-medial relationship between the poem/island body and the visual hologram and then conclude by considering the additional layer of music as an anti-aesthetic.

Sullom is a short filmpoem running 2 minutes 17 seconds that captures collective memories of Sullom Voe through both image and poetry. The memories could be from the speaker of the poem or the larger Shetland community. The filmpoem functions both individually and collectively. *Sullom* contains three primary elements examined in the rest of the chapter: poetry read through voice-over; film with sustained establishing shots of Sullom Voe from a distance; and meandering, nostalgic piano music playing throughout. The visual elements of the film mirror the imagery illustrated in the poem. The film portion shows the landscape adjacent to Sullom Voe and distant shots of the oil terminal, with fiery gas flares, smokestacks burning, and wind blown clouds across the horizon. The film incorporates shots during both day and night, a technique capturing the intensity of light produced by the terminal at night.

Figure 5.1 'Sullom Voe on Horizon'. *Sullom* (Roseanne Watt, 2014), with kind permissions from Watt.

The sense of distance in *Sullom* is not only pragmatic – that is, strict security and privacy laws surrounding oil terminals make it impossible to film at a closer distance (see Figure 5.1). It is also purposeful. The blurring of distance and proximity underscore the filmpoem's objective, which is to capture Sullom Voe from the perspective of a Shetlander, both a universal presence and unknown site of secrecy and protection. From the perspective of *Sullom*, the oil terminal and surrounding environs appear as a memory and real place, even though the real feels more imagined because of limited access to the refinery. Anonymity remains central to petrochemical identity. The film portion of *Sullom* captures the elusive, dreamlike quality of Sullom Voe through the visual medium while the poem describes the personal feelings of the speaker.

The first part of the poem identifies the image of a body as an overt metaphor of the island as a human body gradually dying and being sucked dry of its life source. The poem begins by juxtaposing human and nonhuman geographies/ bodies as a petrospace:

> And here lies
> the slow black pulse
> of the islands,
> which grew roads
> arterial, capillaries across the map;

The multifaceted body here reflects the character of oil as both an abstract idea and a material reality. It refers to human bodies (with anatomical descriptions in the poem) and geographical bodies (across the 'map' of the 'islands'). The poetic form is also considered a literary body, one taxonomy of literary production but one condensed into a self-containing unit. The real and imagined bodies assume a process of decay, speculating about a post-oil landscape in Britain and what that might mean for Shetlanders. This effect generates feelings of solastalgia, where the speaker of the poem recognises desolation of place through ecological and social deterioration, which demonstrates a difficulty to derive solace from her home environment.

Watt describes that her filmpoem project originated as a link between self and place; it captures the visceral experience of the land, particularly while living in Shetland's 'golden era' of oil. Some might argue that oil booms in other Western petrospaces like Shetland – such as St. John's, Newfoundland and Labrador, Aberdeen, Scotland, and Fort McMurray, Alberta – brought with it robust economies and culture, along with a diversity of people who sought to revitalise urban centres (House 2003, 15). Boom-bust economies may advantage communities for short periods, but the long-term effect is dire (hence the 'bust' end of the binary). Home prices increase along with inflated oil revenues, but rarely drop when oil revenues fall. Basic needs, such as food, social services, and transportation, rise without the resources to support them. Crime rates increase because of influxes of transient labourers seeking entertainment when not working. Women and minority groups are less safe because they are targets of exploitation in a society of wealth, excess, and deregulation.

All of these circumstances of petrospaces, among others, create forms of inequality between global capital – generated for multinational corporations from local areas – and local currency. Energy-producing societies also produce political corruption and increased police and military presence to protect oil interests, while at the same time controlling protests and resistance (Urry 2013, 15). Although Shetland may be a smaller area, and possibly one less affected by such imbalances as other larger petrostates, there is dissonance between the values and perceptions that exist between interests of multinational petrochemical companies and the needs of local communities and the subjective experiences of people within them.

Sullom discloses the subjective experience of local voices of those living in this place, but it does so through a time-space expansion of memory. The underlying element in *Sullom* that invokes the corporeal effect is the affective relationship between human and geographical bodies through experience, perception, and values. As mentioned in Chapter 1, the human geographer Yi-Fu Tuan assessed humans' sensorial relationship to space and place through what he called 'topophilia', or 'love of place', where there exists an 'affective bond between people and place or setting' (1974, 136). In a similar way, it would be productive to acknowledge the process of petrophilia – or, an affective bond that has developed between people and oil (drawing on Tuan's spatial term for place-attachment) – literally meaning 'love for oil'. Love here could also serve as addiction or dependence based upon the ways in which oil shapes our lives, thereby creating what Brett Bloom calls a 'petro-subjectivity', where human subjectivity in contemporary society relies on oil in all ways of living on a daily basis (2015, 19). The poem describes the 'slick spill of dark / spinal fluid right / down the middle' of the island. The affective relationship between oil and people, or petrophilia, appearing in these lines compares the image of oil to one of the human and island body. Images of oil extraction aligned with poetic lines of dying bodies create a sense of domination and destruction, and provokes an implicit call to change the energy and ecological future of Shetland.

This multifaceted body is also part of the poem's formal structure. When reading the text, the words are constructed narrowly 'down the middle' of the 'spinal' page (see the URL in note 1 to access the entire poem). Even the staccato constant sounds of the words 'slick' and 'dark' create a muscular physicality, as well as the slippery alliteration – 'slick spill of . . . spinal fluid' – evoking the viscous and elastic qualities of oil. In this sense, it remains difficult to separate the body/oil link even through the structure of the poem. *Sullom* provides a literary map to distinguish a relationship between the past and present, body and oil, refinery and island.

Literary cartography allows us to examine mapped discourses by uncovering literary narrative strategies; it provides a method to read a landscape as both a product of real (topography) spaces and imagined (memory) places. When examining experiential representations of mapped spaces, the poetic style and organisational patterning in *Sullom* characterise how memory and place consistently question how well one really knows history, a place, and so ultimately one's relationship to them. Self-reflective poetry in this case is a method of mapping the island, but also oneself and one's link to the 'memory map', where the processes of memory, identification, and sense of place are integral to the physical and cultural topographies.

In this sense, then, bodies are mapped in the poem through memory. Memory mapping is a way to record subjective memories tied to specific political geographies. The geographer Joan Schwartz explains that 'memory maps' do not describe the real topographical contours of the land, but are 'cartographic expressions of a sense of place', mediated, reinforced, and shaped by memory (1991, 13).[3] Schwartz continues to explain how memory

> maps are neither detailed nor accurate. The spatial relations they communicate are more idiosyncratic than cartographic. The maps reveal more about dynamics than distances, more about the geographical imagination than topographical reality, more about identity than genealogy, more about the character of memory than the nature of land. (1991, 11)

The speaker of the poem in *Sullom* describes her experience of the island/personal body through a process of mapping both as a literary and visual structure, revealing 'more about dynamics than distances'. In *Sullom*, the memory map describes the subjective self as it is mediated through space and memory in a specific petro-history. The speaker recognises the link between place and home in Shetland, described and experienced as 'city of the night / tucked in a far corner / of my north'. The multimedia map does not detail the terrain of certain topographies in technical cartographic terms. Instead, it illuminates issues of the past by way of memory that relate to the present through identity and the cultural impulse of a certain place. The 'far corner' of the speaker's memory parallels the geographical location in the North Sea. But, at the same time, the 'city of the night' of Sullom Voe continues to refine oil.

The second part of the poem moves from the physical or real body to map memory via the speaker's imagination of Shetland. Memory and perception also contain spatial features because they represent ways of being in the world through a sense of time and space in both the present and past for the speaker of the poem. Simply put, there is simultaneity of time in space for the speaker, an effect equally reflected in the visual structure of the film. The history of oil is a history unfolding into the current energy transition and future. But unlike the time-space compression occurrence of modernity, the filmpoem creates space for expansion through the slowness of the images, lingering of words spoken, and stillness of sustained landscape shots. *Sullom* achieves this affect through the contrast between visual and poetic mediums. The poem ends with the memories of the speaker's 'north', and sense of place, when she describes it as 'my north', embodying the now dislocated personal and geographical bodies. The speaker's 'north' is both spatial and historical, but designed to reclaim the narrative of Shetland within a petromodernity through visual images interlinked with poetic text.

The speaker concludes with the two lines: 'a hologram, / a mirknin [darkening] heart'. George Mackay Brown's words drawn from his poem 'Scapa Flow' appear in the opening of *Sullom*: 'And soon the veins of oil will ebb and flow'.[4]

Figure 5.2 'Dead Gull'. *Sullom* (Roseanne Watt, 2014), with kind permissions from Watt.

They also invoke the corporeal aspect of both oil and islands with darkening hearts. What are the Shetland people left with once this 'darkening heart' of the oil industry reaches its inevitable conclusion, when the 'veins of oil' are drained from the body of the islands and sea? One possibility might be represented by the dead gull in the foreground toward the lower right-hand corner of the screen as the lines 'darkening heart' are spoken with an out of focus image of Sullom Voe contrasted in the background (see Figure 5.2). However, the 'hologram' mentioned at the end of the poem provides another vital perspective.

Figure 5.3 'Flickering World'. *Sullom* (Roseanne Watt, 2014), with kind permissions from Watt.

A hologram is a 3D multidimensional light field. Holograms are not only a real image. They are also imaginative – as an interference of the light field, increasing depth and perspective, while also creating confusion and uncertainty for a viewer similar to that of a memory. The poem creates a holographic memory through words:

> In my memory
> it is lit
> as a magic lantern,
> all chatoyant glitter
> of another, flickering world

The audio of the speaker reading the poem combines with the visual element to complicate the viewer's line of vision and perception. The film distorts in and out through deep and shallow focus, creating both a sense of remoteness and closeness between holograms of 'another flickering world' of light and the background oil site of Sullom Voe (see Figure 5.3).

At the same time, the phrase 'chatoyant glitter' – which is a band of bright reflected light created by various angles in a gemstone – reinforces the imaginative and literal effects of perception for the reader and his/her audience of the filmpoem. The visual and poetic holography illuminate other dimensions of the light field, as well as absences, paralleling the petrospaces in a place such as Sullom Voe. The filmpoem is simultaneously material and abstract, physical and symbolic – all of which conjure a state of solastalgia because of how it creates a loss or dislocation, while at the same time invoking a sense of being in the same place.

As the filmpoem demonstrates, there are gas flares burning endlessly, which can be seen (see Figure 5.4). The oil retrieved from the ground is refined enough to use as a fuel and energy resource – a process only imagined and not seen

Figure 5.4 'Gas Flare'. *Sullom* (Roseanne Watt, 2014), with kind permissions from Watt.

in the filmpoem. Images of oil conjure both the social and spatial, while they also allude to environmental damage. The real and imagined images of endless oil extraction ultimately gesture toward destruction because this process could not remain sustainable indefinitely. Images of oil and gas refineries capture the petrochemical-funded industrial landscapes, with flares and smoke interspersed among steel pipes and concrete containers. Such post-apocalyptic images do more than eliminate any sense of place. They spotlight a world void of life.

Anti-aesthetic effect of oil

An additional element of *Sullom* is the musical score – a sequence of piano scales in a natural minor key played throughout the filmpoem. The audio layer provides another real and imagined quality, where the real piano creates an antithetical feeling in contrast to the difficult ideas raised in the poem about black pulses and darkening hearts. The piano music conjures pathos, but this emotional appeal might not strike the right chord for the viewer because the music fails to aesthetically match the poignancy of the filmpoem.

If we look at it another way, especially through an imagined state of reflection, then the film in combination with the music mirrors recent advertising campaigns put out by petrochemical companies designed to rebrand with a friendlier association between oil consumption and society. The rebranding efforts deployed by petroleum companies primarily avoid direct usage of the words oil or gas, both of which are perceived as dirty. For example, Shell Oil has dropped 'Oil' from their name. Petrochemical companies have started replacing any connection with crude oil in their names or image. Instead, they use progressive social associations of energy, such as visions of opportunity and growth, as well as possible ways of living in the world that appears free of conflict. Enbridge Inc. launched a campaign simply called 'E=' (energy equals), emphasising the progressive idea of energy rather than polluting associations with fossil fuels. The various advertisements display images that range from families and animals to charming images of bucolic landscapes. These images replace the actual consequences that the product they produce and sell has on global air and water quality.

In the autumn of 2013, Suncor Energy Inc. (Canada's largest energy company) launched its first major media campaign entitled *See What Yes Can Do*. This advertisement, like others from Shell, Enbridge Inc., or Cenovus Inc., does not appear to promote oil or gas as much as it endorses the cultural ideals of those who consume it.[5] The advertisement basically argues that the petroculture in which we live remains integral to everything people cherish by claiming Suncor produces 'the energy that makes our lives easier'.

The Suncor advert begins by showing images of forests nestled into snow-capped mountains, which are juxtaposed against accelerating images of clouds moving in the sky. The film technique of speeding up images by creating more frames per second in the video effectively builds credibility for the subject being sold (i.e. Suncor as a company) because it establishes a sense of trust moving into the future. The video then cuts to aerial views of city lights and then to cars driving on roads through city blocks, where the speed of the film has again increased, but this time to confuse time and space in a wider world of energy. Using simple film cuts from one extreme image to another, each cut presents an idea of 'change', of which energy is the catalyst – from how we eat, to how we are educated, to how we 'further our creativity' as the advert claims. The video also exhibits scientists working in a lab that demonstrates an appeal to ethos as a way to boost credibility or authority related to Suncor Energy Inc. A closing line of the advertisement asks: 'is there a better way to get things done?' It then affirms, 'yes', so 'come on and see what *yes* can do'.

Throughout the Suncor advertisement, and many others like it, the accompanying music resembles the music in *Sullom*. Melodious piano appears waxing and waning in and out of the background, eventually building to a climax toward the end. As a general rule, musical scores of petrochemical company advertisements present a generic Walmart version of Philip Glass's famous film scores, attempting something cinematic but sacrificing weighty darkness for sweet superficiality. The scores in oil adverts strike a positive note for audiences rather than a cacophony of notes – which is characteristic of Glass and other film scores like his that centralise the piano often with accompanied stringed instruments, such as the cello or violin.

Advertising uses music to subtly draw attention to the product, brand, or idea through emotion (or pathos). Appealing to pathos, in addition to ethos drawing on the company's credibility, is another attempt at sentimentality to persuade the viewer/consumer toward a particular position. With Suncor, among other energy companies, the use of music subtlety underscores the emotional importance of oil and gas in society. Music that aligns with this aim would be slow and flowing, such as simple chord progressions on a piano with consistent foot pedalling to provide echo and to fill sound-space in the videos. The aim of employing music in advertising is to elicit positive emotion correlated with the product or concept being sold.

Sullom is not an advertisement, however. It is a piece of visual and artistic culture. For my purposes here in this chapter, it is an artistic appeal to understand the social and spatial dynamics of a complex and contested petrohistory in Shetland. The chord progression of the piano music used in *Sullom* begins delicately and yet continues to climb throughout the song. The right hand of the performer floats around the diatonic notes in the minor scale, while the left hand plays a simple syncopated pattern, which remains true to the same pitches. Another name for such a scale is the Aeolian mode, often associated with feelings of sadness, melancholy, and deep introspection. This musical effect in *Sullom* also creates a mood of sentimentality correlated with nostalgia, but we might also call it solastalgia, with melancholy linked to deep introspection and sadness related to a loss of place-home for the speaker of the filmpoem. What does a music score of loss of place sound like? What is the soundtrack of desolation? Would it not be similar to what we associate with solastalgia?

In terms of space, the use of solo piano creates gaps in the music, affording the speaker more space to recite the poem in *Sullom*. A similar effect occurs with the confident male voice-over in the Suncor advert. In it, he explains with a firm articulated voice the opportunities of energy under Suncor, a world of 'yes' that confirms action, certainty, and movement. But rather than the underlying melancholy of the speaker in *Sullom*, the Suncor advertisement uses a speaker with firm assurance of our hopeful future. The images in *Sullom* seem to resemble energy advertisements more generally, cutting from natural images of mountains and ocean inlets to lights during the dusk in oil refineries, but the music in *Sullom* adds another layer of melancholia. *Sullom* both gestures toward energy advertisements, while complicating them by striking a deeper mood, conjuring melancholia rather than hope.

The music in the video portion of *Sullom* is an ironic gesture toward, not an attempt to copy, the music used in oil and gas advertisements. What Hal Foster has called 'anti-aesthetic' signals when a critique 'destructures the order of representations in order to reinscribe' the original aesthetic (1983, xv–xvi). Foster explains that an '"anti-aesthetic" marks a cultural position on the present', challenging traditional notions of the aesthetic practice outside of political interpretation 'without "purpose"', all but beyond history, or that art can now affect a world at once (inter)subjective' as 'a symbolic totality'. The anti-aesthetic 'practice' signals a 'cross-disciplinary' approach that is 'sensitive to cultural forms engaged in a politic' denying an otherwise 'privileged aesthetic realm' (Foster 1983, xv). Putting all this together, the anti-aesthetic reframes a cultural position on the present, while critiquing other forms of the social order through politics and space. By using an anti-aesthetic quality to the music, *Sullom* exhibits a transdisciplinary approach that confronts the petrospatial form 'engaged in a politic'. In this way, the music in *Sullom* offers an anti-aesthetic overlay to the bi-medial filmpoem, thereby providing a rich and complex unity of not only media and literature, but also of the petroculture in which it is situated in and largely responding to.

Watt's *Sullom* reclaims the dominant medium of energy advertisements by mimicking the music, but also complicating it through the use of an Aeolian mode, destructuring the order of energy media representations through a re-inscription of the musical score. This effect requires that the words spoken in the poem carry the burden of deeper meaning, which is ultimately where we turn when deciphering the context of the poem. Once the anti-aesthetic application of the music appears in contrast to the visual text, the poem then emerges as the focus. It jumps out at the reader amid the background music subtly ironising aesthetic attempts in energy advertisements. The music illuminates the bi-medial relationship of visual and poetic text by using a similar but more complex version of scores in energy advertisements.

Immemorial places of action

The Scottish novelist George Mackay Brown, who was from the Orkney Islands and wrote about North Sea oil in his 1972 novel *Greenvoe*, famously claimed

Shetland remains haunted by time. As this chapter argues, Shetland is also haunted by space, particularly through displacement and development throughout its own petromodernity. Like other Shetlander poets before her, such as Christine De Luca and Christie Williamson, Watt contributes literary and visual responses to a long and unsettled history of exploitation and cultural transformation through Norwegian colonisation in the Medieval period, then land clearing and displacement in the eighteenth and nineteenth centuries, to oil exploration and exploitation in the twentieth and twenty-first centuries. The body of the poem is itself part of Shetland history, perhaps one that is also dying as well, as many people are moving to more urban areas and off of the islands. The vacuum left by emigration remains difficult to fill through arts and culture when there is no longer a population to support it.

In contrast to Brown's observation, Watt establishes an immemorial portrait through visual culture and poetry of Shetland. She grounds this collective memory in space-time of the past and present not as a historical, nostalgic reverie elevating her locality or community, but as a solastalgic picture of what remains of place surrounding Sullom Voe. David Harvey similarly observes of the affective relationship to place articulated in Gaston Bachelard's intimate spaces: 'And if it is true that time is always memorialized not as flow, but as memories of experiences places and spaces, then history must indeed give way to poetry, time to space, as the fundamental material and social expression' (1990, 218). Watt experiences such memories of this place through poetry because it produces new forms of material and social expression moving in the present, while also shifting back into the past.

By interlinking the poetic medium with film and music, Watt highlights the effect of the poem in contrast to these other elements (as in an anti-aesthetic). She also visualises the loss of place for the viewer watching on the Internet as a gesture toward justice and action. As Henri Lefebvre observed, the visual register has become the dominant mode of the senses in the late capitalist mode of production. In doing so, the effect has been to produce an 'abstract space' (Lefebvre 1991, 287–8). By adding a visual element, however, *Sullom* draws a wider audience to the issue surrounding oil and gas in Shetland by relying on the dominant sense of the historical and spatial petromodernity. But due to *Sullom*'s employment of an anti-aesthetic, cleverly ironising promotional advertising campaigns from petrochemical companies, the poem can then be drawn out as the dominant mode of production countering the abstract spaces of late capitalism.

By examining oil as a spatial and cultural issue of environmental injustice, we can conclude that Watt's overlapping of literary and media constructed filmpoem underscores the combination of real and imagined experiences that ultimately reclaim and construct narratives about socially produced spaces of oil and gas in the Shetland Isles (and Scotland more generally). In doing so, Watt subtly illustrates the ecological displacement associated with solastalgia – documenting and feeling the pain of dislocation, while still living in the same place – thereby underscoring oil-producing spaces such as Sullom Voe as an issue of spatial justice. The added benefit of publishing *Sullom* on the open-access website *Documenting Britain* provides an activist element that confronts contemporary and historical issues of North Sea oil and gas from a Shetlander perspective.

Notes

1 To view the website for *Documenting Britain,* see www.documentingbritain.com. To see the filmpoem *Sullom* with the text of the poem, see www.documentingbritain.com/roseannewatt/an-introduction.html. There are other links between literary and media culture in Shetland. *Writing the North* is a similar Web-based project that documents the links between literature and place of the Shetland and Orkney Islands through online media. See www.writingthenorth.com and Fielding and Smith (2014).
2 See, for example, Finlay et al. (2015), which offers literary responses to energy technologies (i.e. marine renewable and tidal energy) off of Orkney.
3 The term 'memory maps' was introduced by the Canadian visual artist and poet Marlene Creates to describe the process of mapping people's experience with land through art.
4 After asking Watt about this quote, she acknowledged her accidental misquote (likely due to oral tradition of repeating this popular phrase in Shetland). She then guided me to the actual quote from Brown: 'And soon the veins of oil will throb and flow' (2006, 461). The change from 'ebb' to 'throb' does not change the meaning and purpose of the quote, but it does accentuate even further the metaphor of the island body illustrated in *Sullom*.
5 Find the link to *See What Yes Can Do* (Part 1) here: www.youtube.com/watch?v=IMvKZKi2mG8. See also the advertisement campaign for Cenovus Energy Inc., which was launched in 2010 as a series of advertisements entitled *More Than Fuel*. Along with persuading audiences that 'Oil is more that just a source of fuel', this second oil advertising campaign also incorporates flowing piano music in the background. See www.youtube.com/watch?v=M4k_8CnH9I0.

Bibliography

Bloom, Brett. 2015. *Petro-Subjectivity: De-Industrializing Our Sense of Self.* Ft. Wayne, IN: Breakdown Press.

Brown, George Mackay. 2006. 'Scapa Flow'. In *The Collected Poems of George Mackay Brown*, edited by Archie Bevan and Brian Murry, 461. London: John Murray.

'Climate Change: Mitigation – Carbon Capture and Storage'. 2007. *Earthwatch Educational Resources, Climate Change* 5.

Fielding, Penny, and Mark Smith, eds. 2014. *Archipelagos: Poems from Writing the North.* Edinburgh: University of Edinburgh Press.

Finlay, Alec, Laura Watts, and Alistair Peebles. 2015. *ebban an' flowan*. Edinburgh: Morning Star.

Foster, Hal. 1983. 'Postmodernism: A Preface'. In *The Anti-Aesthetic: Essays on Postmodern Culture*, edited by Hal Foster, ix–xvi. Port Townsend, WA: Bay Press.

Gladwin, Derek. 2016. *Contentious Terrains: Boglands, Ireland, Postcolonial Gothic.* Cork: Cork University Press.

Harvey, David. 1990. *The Condition of Postmodernity: An Enquiry into the Origins of Cultural Change*. Oxford: Blackwell.

House, J.D. 2003. 'Myths and Realities about Petroleum-Related Development: Lessons for British Columbia from Atlantic Canada and the North Sea'. *Journal of Canadian Studies / Revue d'Études canadiennes* 37(4): 9–32.

Lefebvre, Henri. 1991 [1974]. *The Production of Space*. Translated by Donald Nicholson-Smith. Oxford: Blackwell.

Macdonald, Graeme. 2015. 'Commentary'. In *The Cheviot, the Stage, and the Black, Black Oil* by John McGrath, edited by Graeme Macdonald, 17–80. London: Bloomsbury.

Schwartz, Joan. 1991. 'Constituting Places of Presence: Landscape, Identity and the Geographical Imagination'. Introduction to *Places of Presence: Newfoundland Kin and Ancestral Land, Newfoundland 1989–1991*, by Marlene Creates, 9–17. St. John's, NL: Killick Press.

See What Yes Can Do. 2014. Suncor Energy. YouTube video, 30 sec. Posted 31 January. Accessed 10 October 2016. www.youtube.com/watch?v=IMvKZKi2mG8.

Shetland in Statistics. 2014. Economic Development Unit, Shetland Islands Council. Lerwick: Shetland Times.

Stevens, Paul. 2016. 'Why We Should Beware the Dawn of Cheap Petrol'. *The Guardian*, 29 February. Accessed 26 March 2016. www.theguardian.com/commentisfree/2016/feb/29/beware-cheap-petrol-crash-oil-prices-climate-change-pensions.

Sullom. 2014. Directed by Roseanne Watt. *Documenting Britain*. Accessed 20 October 2015. www.documentingbritain.com/roseannewatt/an-introduction.html.

'Total's Shetland Gas Plant Officially Opened'. 2016. *BBC News*, 16 May. Accessed 11 October 2016. www.bbc.com/news/uk-scotland-north-east-orkney-shetland-36297571.

Tuan, Yi-Fu. 1974. *Topophilia: A Study of Environmental Perception, Attitudes, and Values*. New York: Columbia University Press.

Urry, John. 2013. *Societies Beyond Oil: Oil Dregs and Social Futures*. London: Zed Books.

6 Pipelines of injustice

More 'oil trouble'

Pipelines are some of the most contentious subjects in contemporary politics. Whether it is the Dakota Access Pipeline running through Standing Rock Sioux Territories or the proposed Canada-United States Keystone XL Pipeline, to the Baku-Tblisi-Ceyhan Pipeline in Turkey, the Kazakhstan-China pipeline from Atasu to Alataw, or the Iran-Pakistan-India Pipeline, multinational oil and gas pipelines penetrate and divide countries and communities. Pipelines also demonstrate unreliable track records of environmental and social safety connected to security. And yet, they are continually proposed and built rarely without meaningful consultation with the collective consent of people living in the places most affected.

In addition to discussing the spatial injustices of pipelines more generally, this chapter examines effects of Shell's Corrib Gas Pipeline on a small community in Ireland as this situation is represented in Richard O'Donnell's (*Risteard Ó Domhnaill*) award-winning Irish documentary film *The Pipe* (2010). This documentary offers a competing story to dominant narratives maintaining that oil and gas pipelines produce jobs and boost the economy. The other story is that people do not benefit from pipelines and they create unnecessary injustices. Because the film focuses on the people affected and the place they continue to live, I want to consider the ecologically and socially displacing effects of the pipeline in their everyday lives and in their community. This chapter is as much about the place affected by the Corrib Gas Pipeline in Rossport, Ireland, and the social responses to it, often called socially conscious criticism, as it is about the documentary film itself. *The Pipe* serves as a visual and cultural narrative of resistance, capturing the social process of storytelling as resistance by the people of Rossport. Perspectives of spatial injustice presented thus far in this book examine geographies and social responses, and the textual medium representing them. While documentaries viewed as visual texts offer a formal dynamic between real and imagined or actual and perceived spaces, they also present social and political reactions to real places that are part of the overlaying narrative. As Imre Szeman has argued in the specific case of oil documentaries, they 'reflect and are a source of the social narratives through which we describe oil to ourselves' (2012, 424). With this in mind, I shall begin by outlining some of the contested spatial and ecological issues with pipelines more generally, with some brief illustrative examples where oil resistances have been previously represented in other artistic forms in Nigerian and Irish culture, before moving to a discussion of *The Pipe* as a visual text that reframes the narrative about the Corrib Gas Pipeline from the perspective of those affected.

Besides tankers, trains, or trucks, pipelines offer another solution for transporting oil and gas across or underneath land and over bodies of water. They transport either natural gas or crude oil. One issue with pipelines is that they tend to run parallel to or near areas where people live without enough oversight or protection from accidents. The people most affected are often the poor and marginalised communities living in rural areas with little political representation. The Dakota Access Pipeline (DAPL) serves as one of the most famous examples of pipeline injustice in North America. The original path of the pipeline was supposed to travel 10 miles northeast of Bismarck, North Dakota; however, the city consisting of 91.6 per cent white population protested because of an unsafe environmental assessment (U.S. Census 2016). Because the city of Bismarck had political representation largely due to their socio-economic class and racial majority, it was then rerouted just 150 metres from the Standing Rock Sioux Territories, an area that although not technically on the Territories would affect the drinking water of the Sioux peoples. Despite the serious resistance famously witnessed by much of the world, the DAPL went ahead as planned. It has already spilled 318 litres (84 gallons) of oil on 4 April and 380 litres (100 gallons) of oil in two separate incidents during March 2017 when construction crews were preparing the DAPL for operation (Canadian Press 2017; Nicholson 2017).

This example of spatial injustice regarding pipelines remains a problem because of their sordid environmental records, often leaking oil or gas and polluting the biodiversity of specific ecosystems for both humans and nonhumans alike. Sunoco Logistics is the subsidiary company under its parent Energy Transfer Partners. Sunoco is building the DAPL even though they have a poor safety record – with over 2,013 leaks from their pipeline construction projects since 2010 (the most in the US), spilling 3,406 barrels of crude oil (Hampton 2016). Seepages of oil or gas contaminate ground and surface drinking water (even from longer distances away), as well as lakes and rivers, while exposing human and nonhuman species to toxification and death. Gas explosions are another ecological concern because they ignite fires, in addition to the collateral damage caused by the initial explosion. Third-party violators – corrosion, lack of maintenance, and unpredictable weather systems, such as hurricanes and earthquakes – remain the main factors for pipeline safety (Kandiyoti 2012, 30). Pipeline incidents vary depending upon state oversight policies, but as historical evidence indicates, it is not *if* a pipeline will spill, it is *when* and how much it will contaminate surrounding water supplies affecting habitats for animals and humans.

Oil and gas pipelines are highly susceptible to catastrophic accidents. Problems such as leaks and explosions are inevitably contingent upon which safety measures are enforced and which are not, and depending upon governmental and safety regulations. The rates of accidents increase or decrease depending upon the amount of funding allocated to ensuring safety. But with deregulation of environmental protection laws in North America, and little to no regulatory laws in countries such as Nigeria and Russia, global pipeline incidents continue to occur (Kandiyoti 2012, 32). Despite the elaborate and effective public relations campaigns of petrochemical companies and politicians sponsored to protect corporate oil interests, all of which are attempts to socially produce an unjust space in society, pipelines are dangerously unstable for the ecosystems and communities near their paths.

Pipelines are both local and global, reflecting that the multiscalarity of spatial injustice, which allows for various stories to be told, whether through dominant narratives of power in corporate media outlets or by local forms of social relationships. As much as they are oppressive and unjustly planned and implemented, pipelines are highly spatial, traversing across topographies and territories for hundreds to thousands of kilometres. One of the exogenous effects (viewed from above as territorial or hierarchical power) of spatial injustice involves the public/private space nexus, and the ways in which public and private spaces must be defended against forms of commodification, privatisation, and state interference (Soja 2010, 45–6). Through the phases of planning, construction, and maintenance, pipelines frequently pass through a variety of nation states with differing environmental and social legal ramifications (Kandiyoti 2012, ix). In this way, pipelines are literally transnational; they snake through various geographies with vastly different communities and laws. This transnationality is why pipelines are also spatial and social in the ways they generate geopolitical resistance. Profit circumnavigates the people and places that directly experience and bear the visual, health, and environmental effects from pipelines in various locations, such as the United States, Afghanistan, India, Canada, Nigeria, Russia, Turkey, or Belgium.

As a symptom of the fossil fuel economy more generally, pipelines are contentious in the ways they are built and maintained. As Rafael Kandiyoti states, 'Conflict pipelines seem to have a certain fatal attraction' (2012, vii). In Nigeria, to cite another distinctive geopolitical example, pipelines that are associated with oil and gas extraction and transport have created frequent ecological and human rights violations. The Nigerian government mass executed environmental protesters resisting Shell's operation in Ogoniland, south-east Nigeria (Kandiyoti 2012, 40). One of the victims, Ken Saro-Wiwa, was a recognised global literary figure. Similar to the writers and filmmakers spotlighted in Part II of this book, such as McGrath, Watt, and O'Donnell, Saro-Wiwa used literary responses to confront the injustices from what he called the 'recolonization' of Ogoniland through oil businesses of corrupt governments (1992, 20). In a letter to the novelist William Boyd, Saro-Wiwa once claimed during his detention from fighting against the 'recolonization' of Ogoniland: 'I've used my talents as a writer to enable the Ogoni people to confront their tormentors. I was not able to do it as a politician or a businessman. My writing did it' (Boyd 1995, 53). When the state colludes with the oil industry, then artistic production serves as one influential area where people can resist such oppression.

Saro-Wiwa risked and ultimately lost his life confronting the spatial injustices of pollution, relocation, and financial disenfranchisement in an oil state through 'the power of the pen' (Saro-Wiwa 1995, 1). Since that time in the 1990s, Saro-Wiwa's writings continue to be one of the most significant global examples resisting multinational oil and corrupted governments that support unjust practices in the fight against what is known as 'blood oil'.[1] Although often associated with dictators in oil states, blood oil describes the human (and nonhuman) and ecological rights violations on a global scale in the quest for exploitative power (whether political, corporate, or a combination of both) to extract and refine petrochemicals for massive private profit (Wenar 2016). The Niger Delta poet Nnimmo Bassey describes the concept of blood oil in his poem 'We Thought It Was Oil': 'We see their Shells / Behind military shields / Evil, horrible, gallows called oilrigs / Drilling our souls'. Bassey then returns to the refrain of the poem: 'We thought it was oil / But it was blood' (2002, 14–15).

Saro-Wiwa reflected that his writing made the most impact for change, not his political or business practices. Here, we see the alignment of ecological approaches with artistic production specifically focused on a spatial injustice related to oil. We do not have to look too far to cite other writers who have confronted resource extraction related to social, ecological, and spatial injustices. In *Heart of Darkness* (1900), Joseph Conrad famously showcased the struggle for resources in the Congo, and subsequent mistreatment of environments (trees), nonhumans (ivory tusks from elephants), and humans (Indigenous peoples), describing this resource imperialism as 'the vilest scramble for loot that ever disfigured the history of human conscience and geographical exploration' (1955, 17). The novel ultimately led to some of the most haunting words in Western literary history: 'the horror'.

Shifting from this brief overview of social and ecological effects of resource extraction in Africa confronted through literary texts, and the tensions surrounding oil production and transport, I want to now focus the aim of this chapter on Ireland and pipelines. Although a former British colony and therefore subject to spatial injustices of empire, Ireland does not contain a resonant literary or visual history of a petrocultural resistance associated with oil and gas.[2]

One notable exception is the Irish playwright and actor Donal O'Kelly. He recently produced a play on the theme of 'blood oil' titled *Little Thing, Big Thing*, which was first performed at the Edinburgh Fringe Festival in 2014. The play is set in Ireland, but the play's central plot revolves around a secret microfilm that contained footage of a murder committed in Nigeria over an oil protest. Martha, an Irish nun who witnessed some of the oil injustices in Nigeria, explains:

> Oil trouble. The Scarab Oil Company drilled a lot of wells. Thousands of them. Pipes everywhere. Right past people's houses, you wouldn't believe what they'd try to get away with! Gas burn-off day and night. Like Living in Hell! Wellheads left unplugged, children playing in filthy bilge, getting sick, listless, dreadful! Spillages, of course! Horrible sticky black stuff! Crops destroyed! People short of food! Rivers polluted! The stench of rotting fish! Some of the locals took exception. Quite rightly! (O'Kelly 2014, 25)

Those who 'took exception' were executed, which alludes to the real and imagined fates of people who resist oil development in Nigeria. O'Kelly's play loosely follows the internationally distressing story of Saro-Wiwa and the concept of 'blood oil' specifically through Nigeria's human rights failings to address the contemporary conflicts between the unrepresented citizens and the petrostate controlled by Shell.

O'Kelly wrote and acted in another play prior to *Little Thing, Big Thing* that mythologises and confronts the Corrib Gas Pipeline project. In *Aillilíú Fionnuala* (2012), which was performed but never published, O'Kelly acts in this one-person play as Ambrose Keogh, a Shell executive who is cursed by a swan to 'tell the truth about Shell'. What follows is the harrowing and vivid tale of Willie Corduff's incarceration and subsequent loss of place because of the pipeline. Corduff is also a central figure of *The Pipe*. Like *The Pipe*, O'Kelly's play underscores the injustices that occurred in Rossport, Ireland, through the people most affected.[3]

Similar to McGrath's play *The Cheviot, the Stag, and the Black, Black Oil* (see Chapter 4), O'Kelly's plays could also be considered petrodramas because they too exhibit a history of petrocapitalism staged as an improvisational theatre performance. The plays focus on oil and the ecological and spatial injustices associated with its production (i.e. extraction, refinement, and shipment). They demonstrate, resembling similar themes that come up in *The Pipe*, how the Irish state's support of private oil companies disenfranchises and displaces local populations, undermining the social and environmental sustainability of smaller communities. Maximising profits for privately controlled companies overtakes the basic rights of uncontaminated drinking water and soil. Unlike in Nigeria, Irish citizens were not murdered by private armies funded by the state for protesting the Corrib Gas Pipeline. Instead, people were incarcerated and dispossessed of their land for resisting Shell's gas pipeline.

Print and visual texts associated with the environmental and energy humanities serve as an effective way of confronting narratives about the benefits of pipelines, whether in the form of a play such as *Little Thing, Big Thing* or in a documentary film such as *The Pipe*, because these texts produce archives of petrocultures that document various injustices through artistic production. For example, literary and visual texts have the capacity to capture audiences' values over time through digestible and relatable social circumstances involving people more than constant exposure to news articles. As *LEGO: Everything Is NOT Awesome* proves (see Chapter 2), a viral YouTube video created more resistance to drilling for oil in the Arctic than decades of news or peer-reviewed scientific articles. Such cultural texts visualise and represent pipelines as cultural objects, rather than engineering feats or generators of economic surplus, drawing objects often seen at the peripheries of the world's oil systems to the centre of the cultural narrative.

Shell to Sea

O'Donnell's debut documentary *The Pipe* focuses on pipeline resistance in the village of Rossport located in north-west County Mayo, in the West of Ireland, along the high cliffs of the Erris coastline and Gengad beach off of Broadhaven Bay. The people of Rossport and the surrounding area were distressed because in 2004 Shell E&P Ireland, supported by the Irish state, started construction on what is officially called the Corrib Offshore Gas Pipeline. It has since been built. With the Corrib natural gas field discovery by Enterprise Oil in 1996, the Corrib gas project aimed to extract natural gas deposits off of the coast of County Mayo and pipe it in near Rossport for refining.[4] The Irish state approved the Corrib project and subsequent construction in 2001 and it became functional on 20 December 2015. The most controversial part of the project involved laying 84 km of pipe underwater and leading up to a refinery site on shore. The Corrib Pipeline now traverses ecologically sensitive waterways near Rossport, but during its planning there was no consultation with the community of which it affected the most. *The Pipe* is about their story, the story of pipelines creating spatial injustices, as a narrative or resistance. The ecological and spatial displacement of the pipeline, along with the history of the Corrib project, is explained in this chapter, along with how this fit into the documentary form.

The discovery of natural gas off of this coastline has created one of the most divisive political and social clashes in the contemporary Republic of Ireland. It resembles similar global conflicts between pipeline protestors and state-backed oil companies constructing the pipelines globally (i.e. DAPL, Nigeria, etc.). During the Corrib gas project, farmers lost rights over their fields, fishers could not access their ocean claims, and people were forced to either move from their homes or forever live near the assembly and completion of the pipeline.

In the process of protesting, five local people known as the 'Rossport 5' were jailed for 94 days in 2005 at the explicit request of Shell E&P Ireland.[5] Earlier forms of moderate opposition began as early 1998, but the Rossport 5 incarceration reached a new level of injustice and triggered a national response. The Rossport 5 proclaimed upon their release:

> We remind Shell and their Irish government partner that imprisonments have historically and will always fail as a method to secure the agreement of Irish people. We now call on our supporters to intensify the campaign for the safety of our community and families. The campaign has now begun in earnest. (Connolly and Lynch 2005, 49)

The arrest of these five men was followed over the next few years by more arrests, unlawful surveillance of the community, and one hunger strike by a schoolteacher.

The 'Shell to Sea' campaign, as the resistance emerging from the Rossport 5 incident was dubbed, argued that the natural gas acquired offshore should be processed at sea, not on shore via a dangerous pipeline. This request had legal precedent. The Kinsale Head gas field in south-west Ireland (County Cork) refined gas offshore during the 1970s and 1980s. Shell to Sea originally wanted to address concerns about the safety of the pipeline and hold the Irish state accountable to fulfil its constitutional obligation to defend the rights of Irish citizens against criminal activity perpetrated by Shell. The Corrib Gas Pipeline is known as an upstream pipeline, which carries unprocessed gas, containing a mix of volatile chemicals, while a downstream pipeline carries clean, processed gas for the consumer (Connolly and Lynch 2005, 27). Shell to Sea also demanded that the Irish government renegotiate the 1992 licensing terms under which the Corrib gas field and other oil and gas companies have operated. Under this 1992 business-friendly licensing agreement, Bertie Ahern, then Minister for Finance but later Taoiseach from 1997 to 2008, reduced corporate tax on oil and gas companies from 50 to 25 per cent. The agreement also afforded oil and gas companies minimal regulation and oversight in the process of exploration and drilling. Moreover, the Irish state and its citizens would have to pay for the oil or gas at full market prices, and Ireland in particular would lose control over prices in times of emergency (Connolly and Lynch 2005, 23).

Covering the Shell to Sea protests, *The Pipe* specifically follows the community members as they resist the pipeline construction. The director O'Donnell is one of these community members who lived in Rossport. He worked in the film industry as a camera operator before deciding to respond to the situation by making his debut film *The Pipe*. The documentary focused on Rossport, but it indirectly spotlights other injustices of oil and gas associated with pipelines, as well as the potential environmental costs, through the cultural and social response to oil oppression more generally. The documentary shows how this event has forever changed Rossport and divided the community – one contingent of the community favouring a short-term economic buyout and the other opting to keep their quality of life rather than accepting money. *The Pipe* follows the lives of several community members in the film, consisting of Willie Corduff (farmer and one of the Rossport 5), Pat O'Donnell (fisher), Monica Müller (farmer), and Maura Harrington (teacher and hunger striker). The larger narrative of *The Pipe* breaks into micro-narratives of each of these local people, with differing stories and experiences to reflect a community feel to the film, but all who resisted the process by which Shell had proposed and implemented the Corrib Gas Pipeline.

The documentary begins by creating a connection to the social, showing people in the community interacting with each other but also as individuals. It then sets up the problem: Shell wants to build a pipeline to transport natural gas from an offshore platform to a refinery on shore. In doing so, *The Pipe* displays the problem in action, or *in medias res*, in the midst of the construction and subsequent protests of the Corrib Pipeline. Because of this, there is no definitive conclusion or solution; instead, it frames a moment in time from 2005 to 2009. With the help of the government's support and police forces, Shell's influence overpowers the community's efforts to block the pipeline. The Irish government provides state-funded police and navy personnel to protect Shell from the local protestors. Shell is even granted a restraining order to keep community members from protesting against the pipeline in their own community. The documentary demonstrates the detrimental effect of the pipeline process by focusing on the affected community of Rossport. Throughout the film, *The Pipe* provides a glimpse into the political fights that the community have waged in Dublin and at the EU in Brussels against Shell's unjust operation to displace villagers from their land and community.

The narrative structure of *The Pipe* suggests that it could be defined as a social or realist documentary, but it fails to unpack the larger problem of pipelines (as I do previously in the opening of the chapter). Rather, it frames the issue specific to Rossport and then documents attempted solutions by the community members that in the end fail. Its pedagogical purpose appears informative because it raises awareness of an issue known outside of Ireland, and so other global viewers might relate, but it also uses pathos, appealing to the emotional connection of the community's struggle throughout the film. The film does not advocate protesting the Corrib Gas Pipeline, which is already in process during the filming. It accentuates the issue of oil in the lives of Irish people and smaller communities in order to potentially avoid other future occurrences of unjust representation and subsequent displacement of Irish citizens. The solution offered in *The Pipe* is paradoxically not to prescribe a fix, but to give voice to a community suffering from the consequences of multiple injustices. While this may not affect change for the people of Rossport, it might force others in Ireland or elsewhere to challenge similar spatially unjust systems that have allowed for similar issues to occur.

The structure in the opening few minutes of the film contains multiple voices that are conflicting and contrasting much like a community narrative. As the opening credits appear on screen, aerial shots sequentially cut from scenic coastlines to rocky cliffs and undulating bogland. This opening succession provides what might initially appear as picturesque landscape shots designed to generate an appreciation of beauty for the viewer. By adopting this visual aesthetic, the documentary immediately appeals to the viewer's sense of pathos and recognition because the Irish landscape evokes a familiarity, as a universal image for those in Western society, forging an emotional connection to this place immediately at the onset. Non-diegetic sound is used with stringed instruments and marimbas spaced out enough to evoke a contrasting sense of anxiety and sadness. The opening shots additionally underscore the focus in the film: the transformation of place-home for the people of Rossport. Panoramas of cliffs, bays, and coastal blanket bogland of County Mayo establish pre-pipeline vistas through powerfully seductive imagery. It helps that Mayo contains some of the most scenic coastlines in Ireland, part of the famed west of Ireland, so O'Donnell harnesses this aesthetic at the beginning to foreshadow the tension that highlights the subject of the documentary.

The next succession of frames, only three minutes into the film, shift to protesters being beaten by Garda (Irish police) with batons. The point of view moves from helicopter panoramic establishing shots of bucolic landscapes and coastlines to close proximity *in situ* footage displaying chaotic interaction of people in physical struggle. These subsequent images are captured with a handheld camera at close distance, often blurred and chaotic with jarring movements on the screen, which is a technique employed in realist filmmaking to document a sense of reality literally occurring 'on the ground'. The camera provides a point of view of an objective character, involved, and embedded in action, rather than a distanced or transcendent character as the contrasting panoramic shots suggest (i.e. god's-eye view/godlike). Reality on the ground represents disorder simulated by an unsteady and unpredictable point of view, whereas planned panoramic point of view shots signify constructed or built reality.

The obvious juxtaposition of the scenic views and violence in the first five minutes serves as a metonymy framing the theme of the entire documentary: there is a stark contrast between how Rossport has existed for centuries to how it now functions in the shadow of the Corrib Gas Pipeline and Refinery. The film sets up what eventually leads to solastalgia – which is to say, the people of Rossport are exiles relegated to remain in their home, but their mobility and support network in the community have been compromised through spatial and social displacement. In the background of the main conflict involving Shell and Rossport, *The Pipe* focuses on the community and the eventual erosion of their place-home, while they remain living there to this day, despite the completion of the pipeline and all of the negative impacts it has had on their lives, such as their work, community relationships, and sense of place.

These opening shots create the spatial scale, as much as a sense of place, especially in how the point of view shifts from distant widescreen aerial shots to proximal and visceral images of conflict. The images of Rossport encapsulate the farmers and community members embattled in a local and global war over land and social rights versus control of oil and gas profit from Shell. The opening sequences visually map the spaces of the film where the injustices unfold, drawing the viewer abruptly and almost shockingly into the chaotic world of Rossport. However, one tension with these opening contrasting shots is that they construct binaries between humans and non-built environments, where humans are constructed as violent and non-built environments or 'nature' are constructed as sublime and majestic. This additional layer of contrast serves to problematise the situation in Rossport. It is not just a battle between Shell and the people of Rossport; it is a conflict between people within the community against their own interests. Many support the Corrib Pipeline, despite its ecological and social consequences, while others support only parts of it or none of it at all. In this sense, the constructed visual comparison at the onset of the film mirrors the deep and complex divide in Rossport as a result of the Corrib Pipeline.

Spaces of resistance

People as well as place provide multiple narratives in the film. Three people elucidate the problems associated with loss of place as a result of the pipeline. Pat O'Connell describes how fishing has always been in his family. He remembers back when he was younger, there were two choices: 'Either you went fishing or you emigrated. So, we chose to go fishing'. He exclaims that 'fishing was in my family, in my blood'. Fishing is more than work; it is a way of being in the world for people such as Pat. This sequence foreshadows what later becomes a battle for local fishing rights against Shell's legal monopoly over the sea. Pat refuses to move his boat so Shell can lay pipe on the seabed. He asserts that by law, his crab nets are protected and it is within his rights to be at sea. Regardless, the police arrest him multiple times for obstructing Shell's effort to lay the pipe. The film abruptly ends with this sequence where Pat utters, 'that's the story for now. Come on, Richie [O'Donnell], I've had enough'. Pat offers one story of resistance in *The Pipe*.

Another community member, Monica Müller, explains her connection to place while she walks across bogland with her dog. She explains that 'for the last 29 years I always loved coming up here. Some people go to church to have this peace but I go here. If everything comes to pass, all you will see is huge machinery ploughing through'. Monica goes on to lament that they will 'build a fence to keep people out of our place'. Although Monica and Pat will continue to live in Rossport, they will view these spaces of ecological transformation every day in the future.

We then hear from Willie Corduff while he walks along his farm overlooking the sea. Willie narrates the concerns of safety for the viewer. Exasperatingly, he explains, 'we wouldn't walk down here today if there was a 340-yard pipe. To think they would put something so dangerous that close to your home and your family. I mean, it's a fear than hasn't been there before'. Willie expresses the main concern of the community members, which emphasises the spatial injustices of pipelines: forced construction with little to no consent from multinational petroleum companies and with no investment in the future of these communities. In order to protect their place-home, the villagers of Rossport are left no option but to resist, and *The Pipe* gives voice to this story of protest.

As these three examples show, the documentary magnifies people's responses to the social and ecological effects of the Corrib Pipeline more than it does the specific fight against Shell. Multiple narratives combine into a collective perspective. In doing so, the film establishes a sense of place defined by many, which might be one way of recognising a response or reproduction of collective narratives as a form of realism. In *Spatiality*, Robert T. Tally Jr. maintains that 'Realism can enable a sense of place and of spatial relations, even if it must impose the form-giving or sense-making parameters through narrative or aesthetic means' (2013, 75). *The Pipe*'s additional narratives of place (established at the beginning of the film) and loss of place (in the scenes of protest) contrast two ways of being in the world. The visual text of the documentary produces another artistic space to document the power struggles related to oil and gas pipelines and the social struggles within communities that result from pipeline construction.

The planning and construction of pipelines create pockets of 'micro-geographies of power', where governing political subjects controlled by larger interests oppress smaller communities. Micro-geographies are more than localised geographical areas (Soja 2010, 36). For Michel Foucault, and other cultural geographers who later adopt this approach, power is relational as much as hierarchical. It 'circulates' much like a 'chain' and 'is employed and exercised through a net-like organization' (1980, 96). In order to map out the entanglements of power, Foucault suggests that forms of analysis 'should be concerned with power at its extremities, in its ultimate destinations, with those points where it becomes capillary, that is, in its more regional and local forms and institutions' (1980, 98). In this way, power functions spatially – both physicality and metaphorically, as well as horizontally (community) and vertically (government and business) – and circulates through local spaces in society, such as communities, industry, or social groups. We must then locate and examine webs of power in the extremities of society, such as different localities where micro-geographies are performed, to untangle the dynamic of domination and resistance, which might include bodies, buildings, or media texts such as documentary films (Panelli 2003, 164).

Within micro-geographies exist groups known as 'micro-minorities', which are smaller indigenous groups who often lack constitutional protection or political legitimacy, especially connected to the wealth belonging to them from their land (Nixon 1996; 2011). In addition to Rossport, the Standing Rock Sioux peoples protesting the DAPL or Ogoni peoples who live in the Niger Delta represent global examples. The micro-geographical dynamic often erupts in previously colonised spaces where power vacuums exist. Within these vacuums, massive corporate interests backing energy companies seize opportunities to extract resources amid a weakened state structure.

Much like reactions to colonialism – which function similarly in how empires have controlled local populations around the globe largely through corrupted or co-opted leaders – micro-geographies result from larger networks of power and social control that ultimately create spaces for territorial dispossession, military occupation, and economic exploitation (Soja 2010, 36).[6] Such spaces of power are controlled through 'discourses of truth' that circulate through corporately owned news media, private oil governments, and the public relations networks of oil companies. While *The Pipe* serves as another discourse of truth, it supplies a counter-narrative, reproducing the social spaces of Rossport through a visual text. However, this visual text captures the experience of the people rather than corporate propaganda. Foucault reminds us that we are 'subjected to the production of truth through power and we cannot exercise power except through the production of truth' (1980, 93). If the discourses of truth in spaces of power are controlled and manipulated, then the layers of narrative in documentaries such as *The Pipe* are competing discourses with each other and with dominant discourses of the state, but the difference is they attempt to reclaim the story of Rossport with the intent to possibly reproduce new spaces of change.

As a realist and social documentary, *The Pipe* aims to recover these narratives by giving voice to the community most affected. The dispute in the documentary results from a larger web of power, but underscores the domination/resistance of a micro-geography inhabited by a micro-minority. This spatial effect not only results in human and ecological injustices; it also presents a model for oil-related injustices involving pipelines, fracking wells, or oil exploration around the Earth. For the micro-minority of Rossport, the Corrib Gas Pipeline represents an invasion and trespass in a place that is considered protected and sacred. Willie, one of the Rossport 5 initially jailed for protecting his land and water, describes how Shell initially approached the pipeline project: 'They came telling us what they were going to do. They never asked us at any stage for permission'. He explains, 'they tried to bully us and it didn't work. Shell decided jail time might reduce our desire to fight'.

The Pipe provides a visual text for the micro-minority of Rossport to produce their own truth to combat the 'production of truth through power'. Without the control of media or support from the Irish government, the people's truth could not overcome the web of power produced by Shell. Documentary films such as *The Pipe* serve as one such outlet to provide narrative of resistance for the micro-minorities. When protesting appeared to be the only option for the people of Rossport, after they were territorially dispossessed from their right to land and water, the police enforced the private interests of Shell instead of the public rights of the Irish people who Shell were displacing. Ironically, and what generates such outrage, Ireland's public purse funded the police to protect Shell's corporate interests rather than the personal and public interests of the Irish citizen. A group of protesters yelled 'blood cops, blood money', echoing similar spatial injustices of 'blood oil' in Nigeria. *The Pipe* addresses a similar question previously asked by other global micro-minorities: if the government defends petrochemical companies, then who protects the people? In particular, if the Irish state shields Shell, then who guards the land and water rights of the people of Rossport?

There are social liberties evolving from this debate that resemble dynamics arising from territorial struggles during colonisation, but there are also environmental concerns that were voiced by the community members, not from either the Irish government or Shell. According to EU law, sand must remain undisturbed from the shorelines because seas and beaches are protected. But, from legal loopholes pertaining to extractive industries in Ireland, these laws do not affect Shell. Legal protections of multinational petrochemical companies supersede the legal rights of Irish citizens (and the EU). Prior to the construction of the Corrib Pipeline, the people of Rossport legally challenged Shell at the EU in Brussels. They argued that inserting the Corrib Gas Pipeline through Broadhaven Bay will create more silt, oil, and dirt that negatively impact habitat and food supplies. The pipeline and refinery will destroy the fishing and crabbing in the area because of chemical and noise pollution, and dry up the markets for selling products to Spain and France. All of these concerns have indeed come to pass since the construction of the pipeline. They have directly affected the livelihood of those in Rossport, forever altering their place-home and forcing them to live like exiles while remaining on their land.

Many more ecological dangers exist due to the proximity and function of Shell's extreme energy project off of Broadhaven Bay. Once out of the bay, the proposed pipeline (at the time of filming the documentary) would snake through bogland on the way to the refinery. Willie expresses his shock at the prospect of digging into the bog: 'They're proposing to go through the bog they haven't looked at yet. They are proposing what they don't know'. Building a pipeline through bogland creates even more instability due to the wetland environment. Willie quips, 'ya can't trust em', referring to the variability of bogs. As mentioned in the previous chapter with the construction of an additional section of Sullom Voe, boglands reduce global climate change by acting as carbon sinks. They also house large areas of biodiversity in an ecosystem. According to the Shell to Sea website, over half a million tons of peat (extracted from bogs) were removed near the local drinking water catchment called Carrowmore Lake in order to facilitate the construction of the Ballinaboy Gas Refinery. In addition to the release of CO_2 into the atmosphere, the removal of such a large amount of peat disturbs the aluminium deposits under the bog. This, along with the mismanagement of water in the area, polluted the local drinking water with high amounts of aluminium. The Carrowmore Lake provides drinking water to over 10,000 residents in the area, which is significant for this rural part of Ireland (Connolly and Lynch, 2005; Leonard 2007, 182–3; *Liquid Assets* 2012).

Besides the impact of bog removal and drinking water contamination, waste chemicals from the Ballinaboy Gas Refinery dump into Broadhaven Bay at the mouth of the Sruthwaddacon Estuary. Such untreated waste includes lead, nickel, magnesium, phosphorous, chromium, arsenic, mercury, and the radioactive gas radon (Shell to Sea). Broadhaven Bay is currently a Special Protected Area, Natural Heritage Area, Area of Special Scenic Importance, and Special Area of Conservation (SAC) because of its shallow bay, with intertidal sand flats, reefs, and salt marshes (Leonard 2007, 182). Other protected conservation areas affected by the Corrib Gas Pipeline and Ballinaboy Gas Refinery include Glenamoy Bog, Blacksod Bay, and Pollatomish Bog. Because of its support from the government and open licensing agreements, Shell was able to circumnavigate many environmental laws protecting the bay.

A 2008 study conducted by the Coastal and Marine Resources Centre at University College Cork (UCC) found Broadhaven Bay to be special importance for marine mammals, expressed as having 'a significant habitat for marine mammals and other biota', with whales, dolphins, seals, sharks, and otters, in addition to their food supply, such as plankton and fish. Irish waters more generally contain 24 species of cetacean (whale), serving as one of Europe's most significant habitats for whales, and is a breeding ground for up to 11 of these species. Within Broadhaven Bay, there were up to 111 separate whale sightings in 2008 at the time of the survey. Among the effects of polluted water and increased sea temperatures, anthropogenic sound can affect behavioural changes of whales through habitat displacement, physical trauma from confused navigation, and hearing loss (Coleman et al. 2009, 3, 7, 24).

The Pipe reveals how the Irish government continues to support the Corrib Gas Pipeline project with the necessary resources of the police and navy against the will of many Irish citizens living in Rossport. Such a model of state-supported private industry functions as a major strategy of neoliberalism and subsequently a model for petrocapitalism. Tax money ultimately disenfranchises the people it is supposed to protect, while it subsidises corporate profits. Irish taxpayers were not permitted to receive any royalties from Corrib gas profits, but development costs of the project were allowed to be written off against paying taxes for Shell (Connolly and Lynch 2005, 4). As of November 2016, the Corrib Gas Sales Estimate was around €331 million (tax-free), while Ireland's revenue from the project was only 7 per cent of this total. The Irish government's profits from oil revenues remain one of the lowest in the world (below 30 per cent), whereas Norway received 75 per cent of oil revenues as a state-owned operation (Johnston 2008, 39). Ireland's pro-corporate system for foreign capital over the past two decades is one main reason the oil and gas industry is focusing its sights there, even though it might not be on the social or geographical peripheries as much as Nigeria or the North Sea (Deckard 2016, 165). This equation epitomises a politics of dispossession, but one in which the micro-minorities must stay and live while industry with state support eventually leaves once the resources are extracted.

The point here is that people – particularly micro-minorities in micro-geographies – have to live with the consequences of social and environmental damage. As of January 2016, the Corrib Gas Pipeline and Ballinaboy Gas Refinery together have cost €3.6 billion to construct, and Shell has admitted that this project will only have a 15–20-year lifespan ('Frightening' 2016). The duration of operations for the Corrib Pipeline and Ballinaboy Refinery will likely be shorter than the period of planning and construction. These costs do not even factor in the amount of publicly tax-funded resources that have been used in the past 15 years to suppress public protests and safeguard Shell's interests in Broadhaven Bay. As *The Pipe* shows through various interviews and footage of resistance, the social consequences of fossil fuels create a police state with militarised zones and limited mobility, dispossessing people from their land and rights to clean drinking water.

One of the themes of the film is that the pipeline threatens the community's health and safety, as well as the environment. The documentary underlines the Rossport community's discord about a solution – whether to accept Shell's money and leave Rossport or stay and fight. However, they all agree that it is not about stopping the construction of the pipeline, but about resolving where it might be built. The initial dispute between Rossport and Shell was more about having the ability to socially produce their own spaces rather than have them be oppressively produced for them. With the latter result, they ultimately had no rights over their own social and geographical spaces.

The people of Rossport and the surrounding area are now exiles in their own micro-geography, unable to leave, but continually witnessing the loss of place now occupied by the Corrib Gas Pipeline and Ballinaboy Gas Refinery. Willie reflects about his resistance to the pipeline: after Shell 'tried to bully us . . . then they just turned around said there's only one answer for this crowd: throw them in jail'. The resistance was fuelled by the need to preserve their place-home from ecological devastation. 'We don't want the community destroyed and that's the reason we are standing up', explains Pat, who then states, 'I have a right to be here. I've been here all of my life'.

For Willie and Pat, among others living in Rossport, the changes to their place-home causes 'psychoterratic' illnesses that are defined as 'earth-related' mental illnesses when one's wellbeing or psyche, as Albrecht and his co-authors explain, is 'threatened by the severing of "healthy" links between themselves and their home/territory' (Albrecht et al. 2007, 95). One major cause to the feeling of homelessness in the place where one continues to live is a feeling of powerlessness to influence and substantially change the outcome. The results of solastalgia on the Rossport community stem from a loss of ecosystem health, relating to a loss of place, which induce a sense of powerlessness, despite efforts to use what power they have access to (i.e. the law) in order to prevent environmental destruction (Albrecht et al. 2007, 96). The specific form of dispossession felt by the people of Rossport directly links to the pipeline project and serves as an environmental and spatial issue of injustice.

Protecting the people

The Pipe addresses many issues related to spatial injustice, such as displacement and dispossession, while linking it to environmental concerns associated with solastalgia. It stands out as one of the few visual or literary texts confronting oil and gas exploration and production in contemporary Ireland. As a social oil documentary, *The Pipe* confronts multinational oil through artistic production. In this way, the documentary form highlights micro-geographical issues by focusing on micro-minorities, thereby producing alternative real and imagined spaces for the local people outside of the neoliberal state. Through deindustrialisation and an increasingly globalised economy, many micro-minority voices cannot be heard or have been systematically silenced because they do not have access to socially produced forms of power. Social documentaries, especially directed and produced by someone from within a micro-minority being documented such as O'Donnell, combat forms of spatial inequalities around the world and highlight the effects of solastalgia.

Drawing our attention to oil as a significant part of space and culture neutralises its often-perceived invisibility, despite the violent and political impacts fossil fuels create in society, and highlights various injustices manifesting through oil production, such as pipelines and offshore oil and gas. The visual and conceptual approaches by O'Donnell in the documentary film reach different audiences in Ireland, as well as globally, since many other micro-geographies are subject to similar spatial injustices of oil imperialism. It displays how oil and gas pipelines transcend national borders with economic epicentres far removed from the places in which they operate.

The Pipe ultimately asks: how are people supposed to react when the law prevents them from protecting themselves or their communities? Who protects the people from the massive oil companies with limitless legal and political representation? The core issue of the film, despite the obvious tension surrounding local rights, community fragmentation, and corporate power, is that we all need energy systems to survive. But the ways in which we produce or extract and transport forms of energy remains an issue of spatial justice and, as Part III shows, climate injustice.

Notes

1 The result of what a former Shell scientist called 'the militarisation of commerce' in Nigeria is nothing short of devastating. Beyond the destructive environmental legacy left behind, where local populations are forced to inhabit the destruction of a once vibrant delta, ingest crude oil in their drinking water, and inhale relentless burning gas fumes, there is also a violent legacy of protest and murder. During the 1990s, the Nigerian military and Mobile Police forces, all of which are funded to some extent by Shell, killed over 2,000 Ogoni people by various forms of murder, such as assassinations or burning of villages. This campaign eventually led to the death of Saro-Wiwa, who was framed for a crime he did not commit and subsequently executed ('Nigeria' 1995, 7–25; Rowell 1995, 10; Nixon 2011, 107).

2 Few print and visual texts deal with oil injustice in Ireland. Written by the American author John Michael Riley, the novel *Dream the Dawn* (2014) is loosely based upon the Corrib gas field exploration off the west of Ireland, and Richard O'Donnell's most recent documentary film entitled *Atlantic* (2016) focuses on fishing and oil in the North Atlantic (along with Newfoundland and Norway). While not creative, Lorna Siggins' journalistic non-fiction book entitled *Once Upon a Time in the West: The Corrib Gas Controversy* (2010) extensively outlines the Corrib gas history.

3 O'Kelly and O'Connell draw on a history of political action and civil protest in County Mayo where the Corrib Pipeline was planned and then built. The Land League, which formed in County Mayo in the nineteenth century, mobilised farmers against inequitable land ownership in the latter half of the nineteenth century. Later in the 1960s and 1970s, the Irish Farmers Association (IFA) / Muintir na Tire and 'Save the West' campaigns in County Mayo drew on this social history of protest originated by the Land League (Leonard 2007, 182).

4 The Corrib gas field is a hydrocarbon-rich geological structure that moves along the Atlantic Margin, stretching from the southern coast of Ireland up to the Norwegian continental shelf. The natural gas made of methane is easier to process than crude oil. As a result of its transnational location in the North Atlantic, the Corrib gas field is controlled by a consortium of Shell (45 per cent), the Norwegian state company Statoil (36 per cent), and Marathon (18 per cent). According to sources from the oil and gas industry, the gas field is worth up to €8 billion. It should also be noted that in 1975 Ireland's government decided not to establish an Irish state company to develop oil and gas resources similar to that of the Norwegian state (Connolly and Lynch 2005, 4).

5 The Rossport 5 included Michéal Ó Seighin, Vincent McGrath, Philip McGrath, Brendan Philbin, and Willie Corduff (the only one with a major role in *The Pipe*).

6 The Irish in Rossport resemble other previously cited examples of petrospatial injustices elsewhere in the world, particularly the Global South, such as the Niger Delta. In 2005, Michael D. Higgins, then Teachta Dála (TD) for the Galway West constituency, but currently President of Ireland, recalls a history of Shell's geographically and socially unjust practices: 'Shell speaks of their projects in Africa where the poorest companies have their resources taken from them by colonising multinationals' (Leonard 2007, 183). Perhaps this is why the people of Rossport took their fight to the EU, since the Irish government at the time supported Shell, much like the Nigerian state has supported Shell for over 50 years.

Bibliography

Albrecht, Glenn, Gina-Maree Sartore, Linda Connor, Nick Higginbotham, Sonia Freeman, Brian Kelly, et al. 2007. 'Solastalgia: The Distress Caused by Environmental Change'. *Australasian Psychiatry* 15(1): 95–8.

Atlantic. 2016. Directed by Richard O'Donnell/*Risteard Ó Domhnaill*. St. John's, Canada, and Harstad, Norway: Wreckhouse Productions and Relations 04 Media.

Bassey, Nnimmo. 2002. *We Thought It Was Oil but It Was Blood*. Ibadan: Kraft Books.

Boyd, William. 1995. 'Death of a Writer'. *New Yorker*, 27 November.

Canadian Press. 2017. 'Dakota Access Pipeline Spilled 380 litres of Oil from 2 Leaks in March'. Canadian Broadcasting Corporation, 22 May. Accessed 23 May 2017. www.cbc.ca/news/business/dakota-access-pipeline-1.4126760.

Coleman, Mary, Evelyn Philpott, Mairéad O'Donovan, Hannah Denniston, Laura Walshe, Margaret Haberlin et al. 2009. *Marine Mammal Monitoring in Broadhaven Bay SAC, 2008*. RSK Environment Ltd. Coastal and Marine Resources Centre, University College Cork.

Conrad, Joseph. 1955. *Last Essays*. London: J.M. Dent & Sons.

Connolly, Frank, and Ronan Lynch. 2005. *The Great Corrib Gas Controversy*. Dublin: Centre for Public Inquiry.

Deckard, Sharae. 2016. 'World-Ecology and Ireland: The Neoliberal Ecological Regime'. *Journal of World-System Research* 22(1): 145–76.

Foucault, Michel. 1980. 'Two Lectures'. In *Power/Knowledge: Selective Interviews and Other Writings 1972–1977*, edited by Colin Gordon, 78–108. New York: Pantheon.

'"Frightening" Shell Gas Pipeline Flares Spook Ireland'. 2016. *RT News*, 3 January. Accessed 18 October 2016. www.rt.com/news/327784-ireland-shell-gas-flares.

Hampton, Liz. 2016. 'Sunoco, Behind Protested Dakota Pipeline Tops U.S. Crude Spill Charts'. *Reuters*, 23 September. Accessed 12 December 2016. www.reuters.com/article/us-usa-pipeline-nativeamericans-safety-i-idUSKCN11T1UW.

Johnston, Daniel. 2008. 'Changing Fiscal Landscape'. *Journal of World Energy Law & Business* 1(1): 31–54.

Kandiyoti, Rafail. 2012 [2008]. *Pipelines: Flowing Oil and Crude Politics*. London: I.B. Tauris.

Leonard, Liam. 2007. *The Environmental Movement in Ireland*. New York: Springer.

Liquid Assets: Ireland's Oil and Gas Resources and How They Could Be Managed for the People's Benefit. 2012. Dublin: Dublin Shell to Sea.

Nicholson, Blake. 2017. 'Dakota Access Pipeline Leaked 84 Gallons of Oil in April'. CTV News, 10 May. Accessed 17 May 2017. www.ctvnews.ca/business/dakota-access-pipeline-leaked-84-gallons-of-oil-in-april-1.3407510.

'Nigeria: The Ogoni Crisis – A Case-Study of Military Repression in Southeastern Nigeria'. 1995. *Human Rights Watch/Africa* 7(5): 7–25.

Nixon, Rob. 1996. 'Pipe Dreams: Ken Saro-Wiwa, Environmental Justice, and Micro-Minority Rights'. *Black Renaissance* 1(1): 35–95.

——. 2011. *Slow Violence and the Environmentalism of the Poor*. Cambridge, MA: Harvard University Press.

O'Kelly, Donal. 2014. *Little Thing, Big Thing*. London: Bloomsbury.

Panelli, Ruth. 2003. *Social Geographies: From Difference to Action*. London: Sage.

The Pipe. 2010. Directed by Richard O'Donnell/*Risteard Ó Domhnaill*. Dublin: Scannáin Inbhear, Underground Films, and Riverside Television.

Riley, John Michael. 2014. *Dream the Dawn*. Asheville, NC: Dry Stack Media.

Rowell, Andy. 1995. 'Trouble Flares in the Delta of Death'. *The Guardian*, 8 November: 10.

Saro-Wiwa, Ken. 1992. *Genocide in Nigeria: The Ogani Tragedy*. Port Harcourt, Nigeria: Saros.

——. 1995. 'Prison Letter'. *Mail and Guardian (Johannesburg)*, 11 November: 1.

Shell to Sea. 'Environmental and Pollution Issues'. Accessed 18 October 2016. www.shelltosea.com/content/environment.

Siggins, Lorna. 2010. *Once Upon a Time in the West: The Corrib Gas Controversy*. Dublin: Transworld Ireland.
Soja, Edward. 2010. *Seeking Spatial Justice*. Minneapolis, MN: University of Minnesota Press.
Szeman, Imre. 2012. 'Crude Aesthetics: The Politics of Oil Documentaries'. *Journal of American Studies* 46(2): 423–39.
Tally Jr., Robert T. 2013. *Spatiality*. New York: Routledge.
U.S. Census Bismarck City, North Dakota. 2016. United Stated Census Bureau. Accessed 1 November 2016. www.census.gov/quickfacts/table/PST045215/3807200.
Wenar, Leif. 2016. *Blood Oil: Tyrants, Violence, and the Rules that Run the World*. New York: Oxford University Press.

Part III

Climate

7 Climate injustice

Everything must change

The pioneering Canadian speculative and science fiction author Margaret Atwood recently wrote an article entitled 'It's Not Climate Change – It's Everything Change' (2015). The title summarises the difficulties with climate change and culture. As Atwood indicates, climate change presents a monumental shift because it alters everything by threatening the existence of humans and nonhumans alike. In particular, the consequences of climate change require we transform knowledge and social formations that construct and support the ways fossil fuel forms of energy have been obtained, produced, and utilised. In the article, Atwood predicts two possible visions of the future: (1) we continue to cultivate alternative energy sources and transform infrastructures that support society without fossil fuels; (2) we continue using fossil fuels as the primary source of energy without a detailed plan for energy transition until they vanish, the result of which would provoke global havoc and mayhem. The former future prepares for the inevitable energy transition to a post-carbon society, taking into account the unsustainable effects of climate change. The latter vision suggests a dystopian future where societies have exhausted the main energy source of oil without foresight into massive infrastructural transitions, creating a future that is difficult to conceptualise and one that would likely erupt into mass chaos.

As a writer of dystopian fiction, Atwood has explored themes of injustice resulting from imbalanced forms of power. The hyperbolic juxtaposition between her two visions for the future serves as a warning about how we as a collective global society might address the self-destructive dependency on fossil fuels to command our energy needs. Atwood's point is that *everything* will change with or without fossil fuel dependency – if, that is, we can survive extreme fluctuations in atmospheric temperature over next 40 years. Society's dependency on carbon-based infrastructures directly relates to global climate change. Besides the notion of 'peak oil', or that a finite supply of extractable oil and gas exists in the Earth, experts are now looking at other pressing issues. One of these concerns is 'peak water', or the idea that because of warming climates, available water resources cannot sustain the expanding global population (Urry 2011, 43).

The consequences stemming from carbon-reliant infrastructures are not only social, but also spatial in the ways drastic variations in global climates disproportionately affect poor and marginalised populations.

Other than nuclear proliferation and war, climate change serves as the most existential threat facing human and nonhuman survival around the globe. Climate change, or more literally called global warming, magnifies global issues of spatial injustice. Climate policies must consider scientific and economic ramifications, but in so doing the social and cultural effects must equally be considered. This chapter treats climate change as an issue of spatial in/justice (climate in/justice) because of the effect it has on the planet and how the people who most suffer the consequences of flooding, lack of drinking water, and diminishing food supplies are not the people who cause or maintain the climate crisis. Rather, people who are displaced and dispossessed as a result of climate change, suffering from forms of solastalgia as exiles, experience ecological injustices caused by those propagating policies that increase climate injustices.

The chapters in Part II broadly explored the relationship between space and fossil fuel culture known as petrocultures, specifically through representations of injustices in print and visual texts of drama, poetry, media, and documentary film. As a logical extension of 'petrospaces', Part III details climate spaces of inequality. As Edward Soja maintains, 'the physical geography of the planet is filled with spatially defined environmental injustices, some of which are now being aggravated by the uneven geographical impact of socially produced climate change and global warming' (2010, 31). Similar to the concept of petrocultures, climate change is a spatial issue of ecological injustice and is socially produced by dominant narratives in culture. By discussing current social and cultural issues, it becomes clear that various print and visual texts already confront prescient issues about climate. The goal here is to make discussions about climate justice ubiquitous, to tell the story through artistic modes of production. Climate change is not only something beyond us in the future, but happening now – simultaneously in the past, present, and future, functioning as a spatiotemporal issue.

Serving as an introduction and analysis to Part III, the following pages explain climate change through a combined scientific and social lens while they also discuss political action (and blockages) toward reducing this change. The final section of this chapter examines the Anglo-Danish documentary released in the UK entitled *Village at the End of the World* (2012). Created by the British filmmaker Sarah Gavron and Danish cinematographer/co-director David Katznelson, *Village* reveals the displacing effects of climate change on the Inuit of Greenland. Gavron and Katznelson construct a visual argument to demonstrate unjust circumstances of climate change, but they do so with stories of climate injustice through the voices and experiences of the Inuit. Similar to the last chapter, Gavron and Katznelson use the visual text of documentary as a socially conscious criticism by way of storytelling as a form of resistance. To this end, this chapter provides a framework for the remaining two chapters in Part III that in one instance examines narratives in 'cli-fi' or climate (change) fiction as a form of transcendental homelessness in McEwan's *Solar* (Chapter 8), and in another example surveys how fiction might ironise catastrophes related to climate change in China Miéville's 'Polynia' and Kevin Barry's 'Fjord of Killary' (Chapter 9).

Climate science and global action

Despite isolated forms of political opposition, the causes and effects of climate change are not controversial among scientists, world leaders, health practitioners, and most global citizens, particularly those suffering direct results of warming climates in the Arctic or Global South. In fact, there is almost unanimity on the cause and impact of the global climate crisis, with over 97 per cent of the climate researchers in agreement that the issue remains a significant threat caused by human activity (i.e. anthropogenic). There are dozens of measurable ways to mark climate-related changes, some of which include documenting less oxygen in the air, ocean warming and acidification, cooling and shrinking in the upper atmosphere, and higher levels of carbon dioxide in trees, coral, oceans, and air. Nevertheless, average rising temperatures continue to be the most expedient indicator of change.

The drastic variations in global climate temperatures result from increases in greenhouse gases (GHGs) because of 250 years of industrialisation reliant upon fossil fuel energies of coal, oil, and gas. Global average temperatures have risen over the twentieth century from 0.6°C to 0.9°C; this rate has doubled over the last 50 years since the beginning of the Great Acceleration (Linden 2006; Metz et al. 2007; Stern 2007; 'Global Warming' 2016). Because of the rise of GHGs, climate scientists predict that the Earth is perilously close to moving beyond a 2°C increase in global temperatures as early as 2050. September 2016, for example, was the first time in recorded history the Earth exceeded 400 parts per million (PPM) of carbon dioxide (CO_2) in the atmosphere. Scientists have warned we must remain under 400 PPM, if not 350 PPM, to avoid catastrophic consequences to the atmosphere affecting the climate and oxygen quality for breathing.

According to the 2017 'State of the Climate' report from the National Oceanic and Atmospheric Administration (NOAA), 2016 was the hottest year on record (going back 137 years) and reached a global increase of 1.5°C (up 0.2°C from 2015, the previously hottest year on record). All five years since 2010 have been sequentially the warmest years on record, culminating in 2016. In fact, the NOAA report also claims all of the 16 years in the twenty-first century rank among the 17 warmest years on record (with 1998 being the eighth warmest). As the data currently shows, 2017 will surpass 2016 in this category.

These extraordinarily hot temperatures distress the health of oceans, which affect global climates, food systems, and water supplies for both humans and nonhumans. Ocean warming continues at such a rapid rate that NOAA has had to rescale its ocean heat charts to accurately measure warming from 2014 to 2016 (NOAA 2017).[1] These changes affect greater numbers of sea life because food supplies and habitats cannot survive warmer temperatures. One well-documented example is the vanishing Great Barrier Reef off of the north-east of Australia. If temperatures continue to escalate based upon the current rate of emissions and without any action to reduce the problem, according to the Stern Report on Climate Change, then the atmosphere could reach 2°C by 2035 (double the pre-industrial levels). Scientists predict that temperatures would likely exceed 5°C by the end of the twenty-first century, which is the equivalent to the change of average temperatures since the last ice age (Stern 2007, vi).

The overwhelming scientific consensus is that the threshold for future human and nonhuman survival requires that global average temperatures remain well below 2°C. If the average global temperature increases beyond 2°C, 3 billion people (one-third of the Earth's population) would experience clean water shortages for drinking and producing food. Extreme global water shortages would diminish food supplies mostly for impoverished populations in the Global South living in densely populated urban centres without infrastructures to withstand extreme weather incidents. Many societies already experience flooding and drought because of changing climate patterns. Bill McKibben, the environmental writer and founder of 350.org, frames it this way: 'Hell is breaking loose now, and we're barely past 1 degree. Two degrees will be exponentially tougher – but 3 degrees will be exponentially tougher than that'. Such an increase is based upon what he calls 'a spectrum of damage', which is a way of measuring those most affected on a sliding scale of vulnerability (McKibben 2017).

The science behind climate change is relatively straightforward. Increases in GHGs create a shield of atmospheric gases trapping heat from the sun. This process generates a greenhouse effect for the Earth. Once temperatures progressively increase, sea levels rise, oceans acidify (affecting sea life, food supplies, and interconnected ecosystems), and ice melts at the poles. These three signs are known as the major climate change indicators. The rate of sea levels rising was higher in the twentieth than the nineteenth century (by a margin of 1.5 mm), but with significant elevation of about 3.1 mm occurring closer to the twenty-first century (between 1993 and 2003). The increase of CO_2 affects the chemical pH balance of seawaters, hence the acidifying effect in oceans and seas. Finally, sea ice continues to decrease at alarming rates. During the summer of 2012, scientists verified the lowest volume of sea ice accumulation in recorded history, which was 49 per cent below the 1979 to 2000 average and, more shockingly, 18 per cent less than the 2007 average (see more about polar ice sheets in Chapter 9) (Stocker and Qin 2013). These consequences point to the cause and subsequent solution: limiting GHG emissions reduces global temperatures and the subsequent effects of atmospheric warming.

GHGs contain various compounds, such as carbon dioxide (CO_2), methane (CH_4), nitrous oxide (N_2O), and fluorinated gases, but out of these CO_2 is the largest accelerator toward atmospheric warming. In fact, CO_2 comprises over 80 per cent of GHG levels. CO_2 is also a primary factor for global warming potential (GWP), which is a process by which scientists can measure the time a gas remains in the atmosphere and the amount of energy it absorbs per pound. The more energy absorbed per pound, the more it contributes to warming the Earth. Here is a breakdown of the levels that each of the GHGs contributes to global warming: CO_2 supplies 81 per cent of GHGs, most of which is caused by human activities, and why carbon is the main contributor of climate change; CH_4 adds 11 per cent of GHGs and 60 per cent of this amount is anthropogenic; and N_2O creates 6 per cent of GHGs and 40 per cent of this is anthropogenic. There are also synthetic fluorinated gases containing hydrofluorocarbons, perfluorocarbons, sulphur hexafluoride, and nitrogen trifluoride that account for the remaining 3 per cent of GHGs. These synthetic gases result from industrial processes. Although emitted in smaller amounts than CO_2, synthetic fluorinated gases contain high GWP and have been associated with creating holes in the ozone layer (EPA 2016).

Science serves as a tool for understanding the scale and scope of problems, while also offering practical solutions to solve them, but scientific understanding alone cannot sustain without political will and social support. The 2007 Stern Review on the Economics of Climate Change made the case that societies have the scientific understanding and technical solutions to decrease climate change. Stern argues that new economic models would both decrease climate risks and produce sustainable models for economic growth. The review specifically concludes that if countries do not act on climate change, the overall costs and risks will be equivalent to losing at least 5 per cent of global GDP per year henceforth. If climate risks and impacts increase, which they already have done since the publication of this report in 2007, then the estimated damages would likely increase to at least 20 per cent of GDP per year. In contrast, the estimated costs of funding action by reducing GHGs would limit the global GDP to about 1 per cent each year (Stern 2007, vi). Regardless of the scientific or economic evidence supporting climate action, a cohesive global vision remains difficult to harness through collective national and international support.

Action must be international because climate change is a global and spatial issue. Various efforts have been made to forge commitments from the industrialised countries, which are responsible for dramatic increases in GHGs. However, proposed carbon emission targets as one form of action are difficult to meet even though they are the most important element of the global climate agreements. Some of the major global events aimed at addressing climate change over the past 40 years include: the First World Climate Conference (1979), the Montreal Protocol about the ozone layer (1987), the Intergovernmental Panel on Climate Change by the United Nations (1988), the United National Framework Convention on Climate Change (1992), the Tokyo Protocol (1997), and the Stern Review (2007). The largest global effort to date occurred in Paris in December 2015.

The 2015 Paris Agreement (COP21) on climate change is perhaps the most effective example to date of a scientific, political, and humanitarian cooperative effort to work globally on climate action. This Agreement built upon the United Nations Framework Convention on Climate Change (UNFCCC) in 1992, and similar Agreements already constructed in Copenhagen (2009) and Cancun (2010), to commit to a larger and more aggressive approach at curbing global GHGs (UNFCCC 1992). All of the 187 countries responsible for 97 per cent of global GHGs 'agree' to submit individualised climate pledges to reduce emissions by 2020, with the further goal to reduce them even more by 2030. The primary target is to decrease the global average temperature to pre-industrial levels (well under 2°C) by reducing all forms of GHGs. Each country, depending upon their status of carbon emitters, must deliver transparent findings about the level of emissions as outlined in the 29 Articles of the Agreement through five-year updates of actions taken and results quantified. This collective agreement states in the 29 Articles that nations recognise 'the need for an effective and progressive response to the urgent threat of climate change on the basis of the best available scientific knowledge'. The Agreement acknowledges the need for 'progressive' responses to 'climate change', but it also affirms 'the importance of education, training, public awareness, public participation, public access to information and cooperation at all levels' ('Paris Agreement' 2015; Suh 2015).

The Paris Agreement is more than a legal document signed by each country and their constitutive governments. It is a social and cultural contract made on behalf of planetary citizens yearning to restore the Earth and its capacity to sustain all forms of life. The Agreement itself does not solve climate change. Global policies of each collective nation with the support of its citizens must prioritise political, economic, and environmental strategies to significantly reduce GHGs from businesses, investors, educators, financial institutions, and municipal governments. The developed/industrialised countries involved in the Paris Agreements decided to set up a way to finance climate action. The goal is to raise $100 billion annually until 2025, which would assist vulnerable communities most affected by climate change.

If information on climate change is clear, scientifically irrefutable based upon verifiable evidence, and identifiable in the ways we can prevent it, and there is global support for collective action, then why is this still an issue to resolve? Scientists and journalists continue to sound the alarm, but evidence and figures do not seem to translate into public and private policy and regulation. The 2015 Paris Agreement is one of the most optimistic advances and resolutions for planetary action to date (even without a commitment from the United States since it pulled out in 2017). Fundamental shifts in social formations hinge on the ability to change values and perceptions about spaces we live in and how we can in turn change them, which has become either implicitly or explicitly an aim for many contemporary writers, filmmakers, and artists. The question remains: will every country succeed in their commitment toward climate action? How can the world decarbonise without the support of larger, polluting governments? The science is clear even if acceptance of it is not. Political and industry opponents present difficult blockages to reach climate targets by failing to reduce GHGs and invest in alternative forms of energy production. How, then, do cultural or artistic representations of climate change provide insight?

Climate breakdown and social change

If non-responses to climate change arise from social and political domains, rather than scientific or journalistic fields, then it might be more apt to call climate change, as the journalist George Monbiot does, 'climate breakdown'. Invoking the idea of climate scientifically and socially breaking down, Monbiot places responsibility on humans as opposed to the unsubstantiated argument that climate change is cyclical (2013). The novelist Benjamin Kunkel calls climate change a 'mild term' compared to 'universal carbon pollution' (2017, 22). Timothy Morton, among many others, prefers the term 'global warming' because of the steady increase in global temperatures since the nineteenth century. He believes that the misleading and politically assuaging term climate change places little to no responsibility on human action (Morton 2013, 3). Due to anthropogenic factors, the temperatures around the planet are increasing at alarming rates and so it is logical and strategic to create terms that accurately describe the threat. The language and narratives in which we frame climate matter.

Climate change involves both climate and weather even though they function differently. Climate, as Morton points out, differs from weather because it contains social and psychological dimensions more than biophysical effects. Climate is the cause, whereas weather is the symptom (Morton 2013, 104). The immediacy of weather can be experienced by humans and nonhumans, but because of the warming atmosphere (due to GHGs), variation of the climate is not as obvious even though it affects weather patterns.

In this way, climate appears abstract and conceptual compared to the empirical aspects of weather – discernible in the moment. To gauge the climate crisis only by immediate weather patterns is to miss the overall problem, which must be measured on a broader scale. Understanding climate is the challenge that often thwarts immediate action. How do we explain or conceive of something so immaterial in non-scientific terms? Morton writes:

> as soon as humans know about climate, weather becomes a flimsy, superficial appearance that is a mere local representation of some much larger phenomenon that is strictly invisible. You can't see or smell climate. Given our brains' processing power, we can't even really think about it all that concretely. (2013, 104)

Trying to imagine climate serves as an almost impossible task for many, whereas weather presents more tangible, if not misleading experiences. This is why people who are sceptical or outright deniers of climate change often base their assumptions on the weather where they live. If they experience a cooler year than usual, then the Earth must not be warming. Researchers found that the ratio of daily record high maximum temperatures to record low temperatures in the United States was about 2 to 1 (Meehl et al. 2009). Spread out over time, the Earth's *climate* continues to warm despite of how people experience *weather* in certain locations. Thus, concretising the effects of climate change in specific narratives about climate versus weather presents a clearer picture, and one that can draw on imaginative practices within the arts and humanities to understand changes to the Earth's climate. How do we solidify what we experience or imagine? A similar process occurs when trying to explain the social production of space by rooting it in the actuality of place.

Warming temperatures caused by climate breakdown are also an issue of justice, signalling a social and political breakdown of institutions. Social and infrastructural systems with extended ties to energy industries remain the main obstacle facing climate action (see Chapter 3). As Upton Sinclair memorably wrote, 'It is difficult to get a man to understand something, when his salary depends on his not understanding it' (1994 [1934], 109). But climate breakdown is not only a political and social process (i.e. lack of adequate protections and economic regulations); it is also a spatial process.

Climate change seems too enormous to imagine in real time, and yet it has documentable consequences. It is a real and conceptual issue and hence presents a climate change paradox that challenges immediate action. Global warming is difficult to conceptualise for humans because it transcends spatio-temporal limits as a biophysical threat simultaneously in the present and future. For example, how might a 2°C increase in planetary temperatures or a rise in sea levels affect peoples' everyday lives? How can we conceptualise temperature changes if they do not directly affect us in a given moment and we cannot be certain of how they may impact us in the future? Changes in climate and the subsequent effects on the environment are difficult to comprehend or accept because of their spatial and temporal magnitude. Climate change is everywhere and simultaneously nowhere. Compressing decades of time across the massiveness of the Earth has led scientists to conclude that alterations to global policies and infrastructures have to occur now in order to save us from future annihilation in 2035, 2050, 2100, or whenever it may happen. The future of climate survival is both in the past and now, occupying space in the social realm that transcends linear time.

Acknowledgement of and action against climate change is therefore psychological and based upon beliefs and value systems because it generates fear both individually and collectively. Accepting such an enormous concept on a large scale, especially one that demands we change everything about our social and economic institutions, seems improbable, if not impossible. In *This Changes Everything*: *Capitalism vs. the Climate* (2014), the journalist Naomi Klein acknowledges the difficulties of recognising global warming because of its massive scope coupled with an overwhelming need for change on a planetary scale. Klein appreciates that for many people, 'climate change' is difficult to keep 'in your head for very long' (2014, 4). Many believe that technology will solve the problem or that climate change will not be as bad as we think. Klein explains how we 'engage in this odd form of on-again-off-again ecological amnesia for perfectly rational reasons. We deny because we fear that letting in the full reality of this crisis will change everything. And we are right' (2014, 4). The prospect of changing everything, circling back to Atwood, seems impracticable and inconceivable. The enormity of the issue triggers paralysis and the paralysis induces apathy. What do people do with the prospect of their own annihilation happening in the moment and also in the future? This presents a climate change paradox.

Another climate change contradiction is called the '*Giddens's paradox*' from Anthony Giddens' book *The Politics of Climate Change* (2009, 2; original emphasis). He claims that because the risks of global warming are not visible in real time, people will fail to acknowledge these effects until it becomes tangible on a smaller, quantifiable scale. The paradox is that once global warming becomes immediately apparent 'in the course of day-to-day life', preventative intervention or 'serious action' to stop 'catastrophes' will be too little too late (Giddens 2009, 2). The future remains difficult to conceptualise, as does the spatial dynamic associated with global warming. Mike Hulme, a climate scientist who worked on the Intergovernmental Panel on Climate Change (IPCC), similarly suggests that an over-reliance on science produces 'mega-solutions' to the 'mega-problems' related to climate change. He maintains that thinking about solutions on such a massive scale, which only triggers the paralysis caused by the climate problem to begin with, has created a 'political log-jam of gigantic proportions, one that is not only insoluble, but one that is perhaps beyond our comprehension' (Hulme 2009, 332–3). Elsewhere, Slavoj Žižek calls this paradox 'cynical reason', which entails awareness without action, especially confronted with disaster. Although referring to ideology, not specifically climate change, cynical reason nonetheless effectively describes the situation when people cannot act on what they cannot understand (Žižek 1989, 25). The conceptual implications of the climate crisis, which mean we must on some level accept the scale and impact of it on the Earth, leads to obstacles in addressing the problem. Denial surfaces as a temporary emotional solution to the paralysis of perception.

The problem with so-called 'climate sceptics' – more aptly called 'deniers' because of the overwhelming evidence to prove anthropogenic climate change exists – is that they argue we cannot *see* or *understand* the effects and causes of global climate changes. However, this may be because climate change affects the planet unevenly, impacting areas in the Global South and outlying areas of the Global North differently than other parts of the planet. The poorest countries are the most vulnerable to the effects of increasing global temperatures. Glenn Albrecht sees a direct link between solastalgia, injustice, and climate change. He writes that as

> bad as local and regional negative transformation is, it is the big picture, the Whole Earth, which is now a home under assault. A feeling of global dread asserts itself as the planet heats and our climate gets more hostile and unpredictable. (Albrecht 2012)

The effect of climate assimilation – or accepting a new normal of global warming – can already be understood and experienced on local and regional levels across the Earth. Turning again to Soja, he argues:

> although global warming and climate change have now been conclusively linked to human agency, it is useful to see that such human agency work through the production and reproduction of unfair geographies and global structures of spatial advantage and disadvantage. This calls for political responses at multiple and interacting scales. (2010, 52–3)

The larger difficulty previously discussed in Chapter 3 is that special interests controlling the neoliberal economy refuse to halt fossil fuel production, which is the leading cause of global warming. Ben Fine has labelled this phenomenon 'economic imperialism' (2002, xi), which allows a small portion of people to benefit at the cost of 99 per cent of the global population. The argument for sustaining the current high-carbon economy does not factor the impact of social and cultural levers that affect the market. The rhetoric claiming there is a conflict between economic growth and reducing carbon levels in the atmosphere related to the fossil economy has dominated since the late eighteenth century, what some define as the beginning of the Anthropocene. Žižek has pointed out that climate change uncovers the 'pseudo-problem' or symptom lying at the heart of the capitalist world-system (2010, 334). Andreas Malm, a human ecologist and author of *Fossil Capital*, similarly observes that 'climate change is not so much a surprising reversal of fortunes as a *lifting of the veil* on two centuries of fossil capital' (2016, 393; original emphasis). But decarbonising the economy, especially with political and corporate interests opposing such a move in spite of economic gain for the majority, is necessary even though it might seem improbable. As the aforementioned 2007 Stern Report confirms, decarbonisation will reduce costs associated with climate change and as a result lead to more economic prosperity – up to 20 per cent higher global GDP. Moving to a low-carbon economy also entails shifting to a low-carbon society and culture (Urry 2011, 3).

To this end, economic growth and environmental sustainability do not present an either/or question, but a both/and also simultaneity, and what Soja has called Thirdspace – an adaptable way of thinking about fluctuating ideas, events, or representations, and how these affect material and perceptual approaches to geographical spaces (1996, 3). The spatial, overlapping the social and historical, offers a multidimensional perspective toward understanding the complexity of global warming as more than just an economic or environmental question, but one that must find accord with all approaches, ideas, or representations.

Acting on climate breakdown requires foresight and imaginative thinking, which the humanities cultivate in society. However, calls to climate action in the humanities have ranged in approach and frequency. In a 2014 issue of the flagship ecocritical journal *ISLE: Interdisciplinary Studies in Literature and the Environment* entitled 'Call to Writers', the editor Scott Slovic, along with environmental philosopher and novelist Kathleen Dean Moore, asked that writers 'set aside their ordinary work and step up to do the work of the moment, which is to stop reckless and profligate fossil fuel economy that is causing climate chaos'. The call urges all types of writers to use their powerful artistic craft 'in the streets, in the halls of politics and power' to confront the planetary climate crisis (Dean Moore and Slovic 2014, 5).

Dean Moore and Slovic devoted a special issue to social action in order to acknowledge the power of artistic work in the humanities and academic writing about it, which reproduces more generative narratives to impact social institutions and formations. Both academic and artistic work in the environmental humanities reinforce the urgency of climate action, but do so by altering perceptions through the imagination as much as emphasising the real consequences substantiated by science. Creating texts or images as socially symbolic acts provides ways to reconstruct spatial injustices related to climate. Forms of fiction and film assist our conceptualisation of this immediate and time-lapsed reality; they offer different frames in which to confront and change value systems that might block rational solutions to the climate crisis.

Artistic responses to climate change have been increasing daily for the past several years. To illustrate this, I want to conclude this section by returning to the example mentioned in the Introduction and cited at times throughout the book. *The Guardian* created a series entitled 'Keep it in the Ground: a poem a day', with a subsequent title 'A climate change poem for today', which was curated by the British Poet Laureate Carol Anne Duffy in 2015. The aim was to collect 20 original poems by various British and Irish poets, such as Alice Oswald, Robert Minhinnick, Paul Muldoon, and Paula Meehan, which were then read by actors, including Jeremy Irons, Kelly Macdonald, Ruth Wilson, and Gabriel Byrne, among others. This project is both activist and artistic in focus because it exemplifies how poetry responds to spatial injustices of place. The 'climate change poem for today' offers a series of poetic responses to confront climate injustices through an open-access source for the public. Some of these poems were published previously in print anthologies, but most were written specifically for this project. Such a multimedia installation produces a diverse response, with poets and actors collaborating through an open platform for global dissemination. The goal of this project was to explore what it might be like if we all read a climate change poem each day and how this process might alter how we value or perceive climate change as an issue related to culture and justice in our personal and collective lives.

The first poem of the series is by the project's curator Duffy and entitled 'Parliament'. The poem speaks directly to the British parliament by warning that 'the albatross / telling of Arctic ice / as the cold, hard moon calved from the earth' will appear in the wake of continued fuel extraction from the Earth. Combining the political, social, and poetic, 'Parliament' encapsulates the nonhuman response to humans' destructive actions: 'The gull said: / Where coral was read, now white, dead / under stunned waters'. Although the albatross offers an infamous literary allusion to Samuel Taylor Coleridge's 'The Rime of the Ancient Marnier', also a poem about human hubris and calamitous weather changes as a result, other animals in the poem offer observations about the effects of fossil fuel energy creating 'Stinking seas' with a 'vast plastic soup, thousand miles / wide as long, of petroleum crap'. The stork, magpie, owl, woodpecker, vulture, macaw, hawk, and nightingale all witness instances of 'Mute oceans. Oil like a gag / on the Gulf of Mexico'.

The response in the poem is logical even if it might appear somewhat fantastical. Why would animals not experience such changes? After all, the scope and range of global warming affects every living organism on Earth. The crane admits, 'What I saw – slow thaw / in permafrost broken terrain' with 'methane breath'.[2] Duffy's 'Parliament' offers a brief example here of a poetic response to the British government that decentres the human in discussions about climate change. What if animals could speak to parliament about the daily issues they witness from climate change? It would add trillions of voices calling for action against the destruction of the planet. In this way, the poem also underlines climate as spatial because it affects global populations of nonhumans and environments as much as humans. By understanding climate change as a spatial issue, which the humanities concretise for viewers through its real and imagined qualities, we can understand both the scope and reality of its injustices.

Documenting the melt

In *Village at the End of the World*, British filmmaker Sarah Gavron and Danish co-director/cinematographer David Katznelson document an Inuit fishing community in north-west Greenland that resists the potential end of their self-sustaining existence after the closure of a Royal Greenland fish factory. Here, I want to illustrate how Gavron and Katznelson's film documents the impact of a proposed government relocation programme, which threatens to dislocate the villagers and fracture the Inuit community's way of life. The documentary demonstrates an example of spatial injustice in the North Atlantic largely caused by global warming and subsequent dislocation. It also shows, much like *The Pipe* in Chapter 6, how this visual text is a narrative of resistance through the voices of those affected.

Village, along with *LEGO: Everything Is NOT Awesome* discussed in Chapter 2, are the only examples in this book of visual texts produced in the UK or Ireland but focused on a geography elsewhere (the Arctic in both cases). They are included in this book because they exhibit a key area of the Global North most affected by spatial injustices of climate change. In particular, *Village* reveals specific instances of ecological displacement related to solastalgia, where the Inuit of Greenland must remain living in their place-homes, but without the resources supporting their traditional ways of life that have sustained them for centuries. Gavron and Katznelson lived with their two children in Greenland for a year while filming the

Figure 7.1 'Niaqornat in Winter'. *Village at the End of the World* (Sarah Gavron and David Katznelson, 2012), with kind permissions from Katznelson.

documentary to experience the challenges of living in an isolated region of the Arctic, and to capture the voice of the Inuit through interviews and their struggles existing with less ice and increasingly warmer temperatures.

Despite these challenges, Karl Kristian Kruse, the chief hunter and village mayor, states that the 'way of the Inuit is to struggle with nature and live sustainably from its fruits'. Although focused on the increasingly precarious existence of the Inuit, who are dependent on dense ice pack for mobility to hunt and fish in warming climates, Greenland and the Inuit who live there are what Gavron refers to as the 'heartbeat of the planet'.[3] In this sense, *Village* is a planetary story about the social and environmental effects of climate change through the camera lens of a British filmmaker, but also through the eyes and experiences of the Inuit of Greenland, who symbolise in the documentary the health of the Earth. As the Inuit politician Aqqaluk Lynge explains, 'What happens in the world happens first in the Arctic' (Gurmen 2013). Using the Arctic as a barometer to monitor the effects of global climate change is not only a scientific process, but one, as this documentary validates, with social and cultural proportions. *Village* documents one specific place reflecting universal concerns experienced by other disenfranchised communities 'on the edge of the world' in both the Global North and South.

The documentary focuses on the village of Niaqornat in north-west Greenland. The village is a hunting and fishing settlement that can only be reached by helicopter or boat when the ice loosens in the late spring (see Figures 7.1 and 7.2). *Village* is broken into four parts following the seasons (and translated into English): winter ('the time of darkness'); spring ('the time of ice'); summer ('when the sun shines'); autumn/winter ('the water begins to freeze'). Niaqornat contains no vegetation or running water. In fact, shortages such as fish, work, entertainment, people, and polar bears are now part of the villagers' austere way of life. The villagers only began using electricity in 1988. The village struggles to stay financially solvent in the twenty-first century because climate change is noticeably affecting their ability to hunt and fish, which serves as both their sole profession and livelihood. Karl reflects upon the current financial realities: 'We've been living as hunters in Greenland for many generations and thousands of years. Of course, you can catch enough in terms of food. But now it's hard to earn money and make a living from it'.

With only 59 residents in the village, the government of Greenland has threatened to eliminate their Danish subsidies if the population of Niaqornat falls below 50 people. Without the subsidies, and the monthly visits from a supply ship providing food that can no longer be hunted locally, the villagers cannot survive. The government proposes relocating the people to a town elsewhere in Greenland. However, the villagers do not want to be forced from their ancestral home. To remedy this dilemma, they make a plan to purchase the now defunct Royal Greenland fish factory in Niaqornat and create a fishing co-op. Erneeraq Therkelsen, one of the villagers, maintains that the 'factory is the heart of the village. If the heart isn't working, the village can't work properly'. Besides the fish factory as a central theme, the documentary follows some villagers through the cycle of a year: Lars the teenager who wants to leave Niaqornat to seek opportunities; Karl the hunter

Figure 7.2 'Niaqornat in Summer'. *Village at the End of the World* (Sarah Gavron and David Katznelson, 2012), with kind permissions from Katznelson.

who believes his generation might be the last to live this way of life (e.g. Lars refuses to hunt); Ilannguaq the 'outsider' who is from South Greenland and works with waste disposal; and Annie, who is a village Elder, relays some of the oral traditions and stories of Inuit Shamen.

The film accentuates the contemporary pressures now facing Inuit villages that have existed in Greenland for centuries. Climate change plays a significant role, but so does the government's aim to eradicate the traditional way of life for the Greenland Inuit by displacing them from their traditional territories, thereby eliminating the necessity for public subsidies that support their communities. The government of Greenland (and by extension Denmark) has historically shown reluctance to subsidise the Inuit, but this issue has become more central now that the Inuit cannot survive on their own through a combination of economic, social, and climate factors. As mentioned in Chapter 1, one gap in understanding spatial justice is how specific forms of artistic production respond to or confront forms of geographical inequalities. Spatial approaches to injustice have remained primarily geographical and social without direct correspondence with work in the humanities. Soja, for example, scrutinised the geographical circumstances of space, but culture and art that materialise from and about these spaces also reveal circumstances of spatial injustice.

Documentary films such as *Village* uncover spatial injustices through everyday lives 'lived as a project', to borrow again from Lefebvre (1991, 406–7). The documentary as a visual text reclaims some of the real and symbolic acts to reproduce social spaces from dominant narratives of power. *Village* achieves this aim by providing an international platform for the Inuit to voice their experiences in their threatened place-home. The film critic Bill Nichols argues:

> documentary is not a reproduction of reality, it is a *representation* of the world we already occupy. It stands for a particular view of the world, one we may never have encountered before even if the aspects of the world that is represented are familiar to us. (2001, 20; original emphasis)

Village provides a *representation* of the world, but in doing so it offers ways of interpreting and reconstructing narratives. For example, *Village* is about a small Inuit fishing community in Greenland, while it also alludes to the challenges of other smaller rural communities in the UK, Ireland, or globally who must face rising economic realities along with ecological destruction that stem from globalised capitalist economies.

The pioneer British documentarian John Grierson believed the documentary as a visual text must deal with the economic underpinnings of the film's subject (Ellis and McLane 2005, 113). Gavron similarly claimed in an interview that *Village* is an observational documentary because it reflects a specific way of living rather than offering an argument. Analogous to *The Pipe* in Chapter 6, *Village* implicitly rather than explicitly calls viewers to action. Gavron and Katznelson may not make an overt argument, but the film's '*representation* of the world' clearly displays the economic realities caused by climate change and how they are specifically experienced by the Inuit. And yet, despite the typical trajectory of documentary form, the film provides character development as a rhetorical device of pathos to persuade audiences of the personal as well as social and spatial difficulties the community experiences. It does so through the film's subject of the isolated and affected topographies of Greenland in the Arctic, while also capturing various individuals and their experiences in this place.

The film opens with the following words on the screen: 'For centuries, Greenland's largely Inuit population has lived in remote settlements on the edge of the ice cap. Many of these villages are now struggling to survive'. The film then shifts to the school in the village, translated as 'School of Hope', where the teacher conveys the biblical myth of Noah and the flood to the students. The teacher, Mathias Therkelsen, discusses the rising sea levels during the archetypical flood myth. He then writes a question on the board for the students to answer: 'What is our future?' The contemporary parallel is obvious. The noticeable effects of global warming and the prediction for the future are major concerns for the Inuit. Sea-level change disturbs the economic and social stability of the village, which become central realities for the Inuit children facing an unknown future. Scrutinising the question about their future draws attention to the link between identity and place-home.

This Arctic village directly suffers from industrial practices occurring elsewhere on the Earth that create pollution, carbon, and global warming. Smaller communities such as Niaqornat – on the perceived edges of the world where ice density remains essential for survival – are most affected by glacier melting. Glaciers are the symbolic and material foundation of their existence. They derive their sense of place-home from the ice pack because it sustains their ways of being and knowing. It is also practical. Hunting provides income for the village through the sales of fur while it also supplies food and pelts for warmth. As Elder Annie recalls, 'ice used to be solid and very thick', but now it is not. Without the dense ice pack in the winter, hunters in the community cannot hunt larger game, such as polar bears, whales, and reindeer. Walking or dog-sledding across miles of ice is now too dangerous. At the same time, the villagers cannot use boats to break through the ice. In addition, the polar bears and reindeer cannot migrate as far because of the increasing ice melt. Arctic animals experience less mobility which limits food options. Dense ice pack is necessary to sustain all forms of life in this region throughout the long winter. In this way, the injustices of displaced climate change distress both the humans and nonhumans in the Arctic.

The problems associated with development on a planetary scale have affected many communities similar to Niaqornat. Neoliberal forms of uneven economic development through what Soja has called the '*globalization of injustice and the injustice of globalization*' creates both social and environmental circumstances that affect Indigenous communities (2010, 57; original emphasis). In particular, changing climates and the subsequent effects on marginalised populations serve as two major obstacles for survival and maintaining place-home. The immediate and visual effects of climate breakdown are most visible in less industrialised communities, such as the Inuit in the Global North or the populations in the Global South. And yet, the root of the problem stems from globalised industrial economies that continue unsustainable practices to continue growth through urbanisation on a macro scale. The injustice documented in the film results from indirect uneven development around the planet, affecting non-industrialised communities who do not have economies or infrastructures to reduce the effects of rising sea levels and melting ice, all of which affect their economic base of hunting and fishing and traditional ways of being and knowing. Climate change, water pollution, and

Figure 7.3 'House in Niaqornat'. *Village at the End of the World* (Sarah Gavron and David Katznelson, 2012), with kind permissions from Katznelson.

diminishing food supply all result from social policy in areas around the globe, affirming the assertion that environmental factors are largely socio-spatially produced and result from a globalisation of injustice. The Inuit of Niaqornat face displacement because of spatially produced forms of global warming occurring in one location but affecting another.

The documentary attempts to locate the experience of changing climates *in situ* even though the spectres of the past and the foreseeable futures are equally present. Some of the film edits highlight this experience. The film juxtaposes images of the vast spaces of thinning ice against images of isolated homes in the village (see Figure 7.3). The contrast of scale is stark and pervasive with the 59 villagers living in a dozen houses in north-west Greenland. Scale is often applied in the documentary form to generate affective responses (Szeman 2012, 436). As Karl walks across the ice with a pole to test the firmness (see Figure 7.4), the camera zooms out to reveal the immense distance he must walk to test the safety before bringing out the dog sleds for hunting. Gavron and Katznelson juxtapose these vast establishing shots of the glacial topographies with personal interviews in isolated rooms of the smaller houses in the village. The visual aesthetic of *Village* captures the effect of the villagers' isolated existence in connection to the vastness of the space in which they live. Their place-home contains this inherent contradiction of titanic and secluded, mirroring the spatial qualities of global climates.

The government's plan to relocate Inuit communities serves as another important element underpinning the documentary that relates to notions of space and injustice. One problem is the inhabitants of Niaqornat are economically disadvantaged through globalised currency. Another issue is the villagers have even less ability to survive in urban locations than in their traditional territories.

Figure 7.4 'Karl Testing Ice Density'. *Village at the End of the World* (Sarah Gavron and David Katznelson, 2012), with kind permissions from Katznelson.

Regardless of international currencies and threats of displacement because of the effects of climate change, the documentary reinforces they should not have to leave. Karl, the hunter and mayor, admits that 'Greenland communities want to be independent and want to sell their rich resources'. Karl's goal for the community, which continues generations of tradition, runs contrary to what *Village* captures playing on the radio: 'The Finance Minister wants to reform the way we live here in Greenland. There isn't enough money to keep unsustainable villages alive. We have to look at how we live. Where are we Greenlanders going to live in the future?' Even if they move to a more populated area in Greenland, they will continually face ecological injustices from uneven development across the planet. Forcibly removing and relocating the villagers to larger towns in Greenland does not address the fundamental issue of spatial climate injustice that villages such as Niaqornat continue to experience.

Instead of accepting a proposed move, the villagers of Niaqornat decide to 'reclaim' their independence by purchasing the defunct fish factory in the village from Royal Greenland and develop it as a cooperative. 'As a community', one villager expresses, 'we've started working to reclaim the fish factory'. They were talking with the government in Nuuk about how to purchase the factory and manage it as a co-op business to collectively support the village. Royal Greenland, a fishing company owned by the government of Greenland, initially proposed a price for the factory, but the villagers' resources to purchase it were not enough. To move forward with this plan, each family in the village had to buy a share in the co-op at 2,000 kroner (about £235/$300), which is a significant amount of money considering no one in the village has a currency-based income aside from the meagre subsidies provided by the government. Despite the plan for the Inuit to become mostly self-sustaining again, Royal Greenland (under the government at the time the film was made in 2010–2011) wanted the villagers and others like them in isolated areas of Greenland to move into towns so subsidies might be reduced or even eliminated.

Even if the villagers were to successfully purchase the fish factory, the government of Greenland doubted the factory could financially support the entire village without monthly subsidies. Karl surmises that Royal Greenland's proposition for the villagers to purchase the factory was purposely made at a higher price than the community could afford. He believes the 'message' the government sent 'is that people must move from villages to the towns'. Karl explains that to 'reopen the fish factory is to reopen the heart of the village. But Royal Greenland is frustrating our intentions. The managers in these institutions don't seem to want progress in this village'. In addition to spatial and climate displacement, the community faces literal removal from their generational place-home, a removal of their 'heart' and centre of existence. Karl recalls, 'we're told that if we move to town, we'll be happier'. Another possible motive for the government's desire to relocate the Inuit and close hamlet villages has to do with extractive or carbon colonisation, which echoes themes discussed in Part II of this book. There are significant oil and gas reserves that the government has recently been more vocal about pursuing. The Inuit Circumpolar Council's Greenland wing has stated on record they are 'immensely worried' about Arctic oil and gas exploration off of Greenland. They maintain that Inuit 'hunters', similar to Karl in *Village*, 'are deeply sceptical towards this development' (Bell 2017).

If the Inuit of Niaqornat live in a larger town, then they could lose their traditional way of life and their sense of place-home. But if they remain, they will suffer solastalgia and the consequences of ecologically induced stresses resulting from environmental changes associated with warming climates. Out of the two options, they decide to remain a sustainable village, but where they support each other, not move into a situation where they will remain economically dislocated and unhappy. Despite the stresses of climate change, remaining is the only option. Relocation from this type of climate-induced colonisation will create social breakdown, fragmenting the community by decimating the cultural Inuit ways of being and knowing. Karl speaks on behalf of the community: 'We're resisting this attempt to close down the villages'. Purchasing the fish factory offers a way for the villagers to reclaim their shifting environment and produce for themselves potentially sustainable futures. Karl acknowledges it is a 'struggle for us to defend this community' against all of the factors out of their control. Despite multiple forms of spatial injustice experienced by the villagers, paralleling similar global situations, their fight for climate justice resulted in purchasing the cooperative fish factory on 1 August 2010.

Niaqornat resists removal through creating a co-op and self-sustaining industry that provides enough for the village to survive on independently from government subsidies. They additionally travel to other Inuit villages in Greenland and provide assistance with renewing other defunct fish factories. Much like Niaqornat, other villages are now operating co-op factories to sustain their way of life in their traditional place-home. Climate change will continually affect the Inuit of Greenland, but with fish factories producing revenue for the villages, the people do not have to rely as much on hunting polar bears and reindeer to sell fur, a practice that could disappear due to the rising temperatures of the Earth. The villages recognise in the documentary it is only a matter of time – perhaps only a decade – before the ice melts enough where they cannot survive winters, nor can they successfully navigate across vast expanses of ice to fish or hunt. In the meantime, they can survive on Greenland's main economic industry: fishing and fish exports.

The Earth's barometer

Gavron and Katznelson's choice of title, *Village at the End of the World*, contains two meanings that serve as a synergistic way to bring this chapter to a close. The first is that the Inuit community of Greenland has existed for generations on the global peripheries, close to the perceived edge of the world. The second implies that the world itself is currently ending, and not only for remote communities such as Niaqornat. The end of the world affects everyone on the planet, which is why *Village* might seem tangential and an isolated case but is globally relevant to everyone anywhere. The title's lack of specificity is purposeful: it is a metonym for the entire planet, which is the 'village' experiencing 'the end of the world' because of diminishing food and water supplies, disappearing professional opportunities in rural areas because of globalisation, and loss of place-home. Even if the Inuit of Niaqornat were to have been forcibly removed to other towns in Greenland, they would continue to experience the ecological consequences of warming climates and melting ice in the places where they live. The entire Arctic is under threat. Their dispossession would produce solastalgia anywhere in Greenland because they would have to bear witness to the climate-related changes to their environments. Returning to the opening quote from the Inuit politician, what happens in the world begins in the Arctic. *Village*, particularly through the Inuit's experiences in the documentary, serves as an acute barometer of the Earth.

The larger point for the chapter is that the temporal dimension of climate change demands, drawing on Atwood's two-part scenario at the beginning of the chapter, that we address our future right now – simultaneously informed by the past and future. The spatial dimension of this question heightens the immediacy of addressing our environmental and climate futures, which are more pressing for disproportionately dislocated communities such as Niaqornat on the 'edges' of the Earth. The spatio-temporal effects of climate change might be gauged through what McKibben (2017) calls 'a spectrum of damage', where the measure of concern or injury is based upon the location and relative global wealth. This is why, in part, the effects of global warming, as well as other planetary injustices, are characterised geographically in the Global North or South, with the latter, except for the Arctic, experiencing greater levels of ecological damage.

Fortunately, the story captured in *Village* offers hope, if only temporarily, of how communities might resist spatial injustices. The villagers of Niaqornat ultimately achieve justice through their own successful use of the resources available to them. In doing so, they resist forced displacement by creating a co-op and self-sustaining industry that provides enough resources for the village to survive independently from the government subsidies. As a visual text, the documentary form and subject serves as a 'socially symbolic act', to return to Jameson (1981), that we might apply elsewhere and through a spatial as well as a historical lens. *Village* serves as a compelling model of spatial injustice in the North Atlantic environmental humanities because of its subject and its cultural and socially engaged documentary form, which allows for viewers to understand a community's struggle to survive, providing a prescient microcosm of the larger environmental problems we face as a planet.

Notes

1 It should be noted that all of the data collected on climate change from the United States government websites was during the Obama Administration in 2016, a period that offered some of the most extensive research on climate change records to date. These websites may no longer exist or the data might be altered under the following Trump Administration.
2 See Duffy (2015). The poem was previously published in Duffy (2014).
3 See Gavron's director statement on the film's website: http://villageattheendoftheworld.com/about.php.

Bibliography

'Adoption of the Paris Agreement'. 2016. United National Framework Convention on Climate Change, 12 December. Accessed December 2016. https://unfccc.int/resource/docs/2015/cop21/eng/l09r01.pdf.

Albrecht, Glenn. 2012. 'The Age of Solastalgia'. *The Conversation*, 7 August. Accessed 2 September 2015. http://theconversation.com/the-age-of-solastalgia-8337.

Atwood, Margaret. 2015. 'It's Not Climate Change – It's Everything Change'. *Matter*, 27 July. Accessed 1 August 2015. https://medium.com/matter/it-s-not-climate-change-it-s-everything-change-8fd9aa671804.

Bell, Jim. 2017. 'ICC-Greenland Worries about Potential Oil-Gas Activity in Baffin Bay'. *Nunatsiaq Online*, 13 January. Accessed 20 February 2017. www.nunatsiaqonline.ca/stories/article/65674icc-greenland_worries_about_oil-gas_development_in_baffin_bay/.

Ellis, Jack C., and Betsy A. McLane. 2005. *A New History of Documentary Film*. New York: Continuum.

Dean Moore, Kathleen, and Scott Slovic. 2014. 'A Call to Writers'. *Interdisciplinary Studies in Literature and Environment* 21(1): 5–8.

Duffy, Carol Anne. 2014. *The Bees*. London: Faber & Faber.

——. 2015. 'Parliament'. *The Guardian*, 27 March. Accessed 28 November 2016. www.theguardian.com/environment/2015/mar/27/keep-it-in-the-ground-a-poem-by-carol-ann-duffy.

Environmental Protection Agency (EPA) (United States). 2016. 'Overview of Greenhouse Gases'. Accessed 2 September 2016. www.epa.gov/ghgemissions/overview-greenhouse-gases.

Fine, Ben. 2002. *The World of Consumption*. London: Routledge.

Giddens, Anthony. 2009. *The Politics of Climate Change*. Cambridge: Polity Press.

'Global Warming'. 2016. National Aeronautics and Space Administration (NASA). Earth Observatory. Accessed 12 December 2016. http://earthobservatory.nasa.gov/Features/GlobalWarming/page2.php.

Gurmen, Esra. 2013. 'The Loneliness of the Village at the End of the World'. VICE, 22 May. Accessed 23 September 2015. www.vice.com/en_us/article/village-at-the-end-of-the-world-sarah-gavron-interview.

Hulme, Mike. 2009. *Why We Disagree about Climate Change: Understanding Controversy, Inaction and Opportunity*. Cambridge: Cambridge University Press.

Jameson, Fredric. 1981. *The Political Unconscious: Narrative as a Socially Symbolic Act*. Ithaca, NY: Cornell University Press.

Klein, Naomi. 2014. *This Changes Everything: Capitalism vs. The Climate*. Toronto: Vintage Canada.

Kunkel, Benjamin. 2017. 'The Capitalocene'. *London Review of Books* 39(5): 22–8.

Lefebvre, Henri. 1991 [1974]. *The Production of Space*. Translated by Donald Nicholson-Smith. Oxford: Blackwell.

Linden, Eugene. 2006. *Winds of Change: Climate, Weather and the Destruction of Civilizations*. New York: Simon & Schuster.

Malm, Andreas. 2016. *Fossil Capital: The Rise of Steam Power and the Roots of Global Warming*. London: Verso.

McKibben, Bill. 2017. 'With the Rise of Trump, Is It Game Over for the Climate Fight?' *Yale Environment 360*, 23 January. Accessed 23 January 2017. http://e360.yale.edu/features/with-the-ascent-of-trump-is-it-game-over-for-the-climate-fight.

Meehl, Gerald A., Claudia Tebaldi, Guy Walton, David Easterling, and Larry McDaniel. 2009. 'Relative Increase of Record High Maximum Temperatures Compared to Record Low Minimum Temperatures in the U.S.' *Geophysical Research Letters*, 1 December. Accessed 4 April 2017. http://onlinelibrary.wiley.com/doi/10.1029/2009GL040736/full.

Metz, Bert, Ogunlade R. Davidson, Peter Bosch, Rutu Dave, and Leo A. Meyer, eds. 2007. *Climate Change 2007*. Fourth Assessment Report of the Intergovernmental Panel on Climate Change (IPCC). New York: Cambridge University Press. Accessed 12 December 2016. www.ipcc.ch/publications_and_data/ar4/wg3/en/contents.html.

Monbiot, George. 2013. 'Climate Change? Try Catastrophic Climate Breakdown'. *The Guardian*, 27 September. Accessed 9 October 2016. www.theguardian.com/environment/georgemonbiot/2013/sep/27/ipcc-climate-change-report-global-warming.

Morton, Timothy. 2013. *Hyperobjects: Philosophy and Ecology after the End of the World*. Minneapolis, MN: University of Minnesota Press.

National Oceanic and Atmospheric Administration (NOAA). 2017. 'State of the Climate: Global Analysis for Annual 2016'. National Centers for Environmental Information, 19 January 2017. Accessed 19 January 2017. www.ncdc.noaa.gov/sotc/global/201613.

Nichols, Bill. 2001. *Introduction to Documentary*. Bloomington, IN: Indiana University Press.

Sinclair, Upton. 1994 [1934]. *I, Candidate for Governor: And How I Got Licked*. Berkeley, CA: University of California Press.

Soja, Edward. 1996. *Thirdspace: Journeys to Los Angeles and Other Real-and-Imagined Places*. Malden, MA: Blackwell.

Soja, Edward. 2010. *Seeking Spatial Justice*. Minneapolis, MN: University of Minnesota Press.

Stern, Nicholas. 2007. *The Economics of Climate Change: The Stern Review*. Cambridge: Cambridge University Press.

Stocker, Thomas F., and Dahe Qin, eds. 2013. *Climate Change 2013: The Physical Science Basis*. Fifth Assessment Report of the Intergovernmental Panel on Climate Change (IPCC). New York: Cambridge University Press. Accessed 12 December 2016. www.ipcc.ch/report/ar5/wg1/.

Suh, Rhea. 2015. 'An Insider's Journal from the Paris Climate Talks'. *Paris 2015: Demand for Climate Action*. Natural Resources Defense Council. Accessed 6 December 2016. https://medium.com/natural-resources-defense-council/an-insider-s-journal-from-the-paris-climate-talks-7cc4d8772962#.337uddb16.

Szeman, Imre. 2012. 'Crude Aesthetics: The Politics of Oil Documentaries'. *Journal of American Studies* 46(2): 423–39.

'United Nations Framework Convention on Climate Change (UNFCCC)'. 1992. United Nations. Accessed 2 December 2016. https://unfccc.int/resource/docs/convkp/conveng.pdf.

Urry, John. 2011. *Climate Change and Society*. Cambridge: Polity Press.

Village at the End of the World. 2012. Directed by Sarah Gavron and David Katznelson. Copenhagen: MET Film.

Žižek, Slavoj. 1989. *The Sublime Object of Ideology*. London: Verso.

——. 2010. *Living in the End Times*. London: Verso.

8 Cli-fi

Hot genre of change

This chapter is about the literary genre or style of writing called 'cli-fi' – otherwise known as climate (change) fiction. Its remit is self-explanatory: fictions focused on anthropogenic climate change as a sociopolitical and environmental issue. The history of cli-fi as a distinctive category is relatively short, with limited critical and popular information about it until about 2010. More recently, cli-fi has gained popular attention on the radio, in bookstores, and in distinguished publications such as *The New Yorker*, *The Guardian*, and the *Chicago Review of Books* (the latter of which has a regular column on cli-fi entitled 'Burning Worlds').

Cli-fi as a recognised literary style is only about 10 years old. Dan Bloom, who is a freelance writer and editor based in Tokyo and Taipei, credits himself for inventing the term back in 2007. Bloom is a self-proclaimed evangelist for climate fiction, and he runs a popular blog, Facebook group, and website called The Cli-Fi Report. The revelation to name it cli-fi appeared to Bloom after reading the 2006 Intergovernmental Panel on Climate Change (IPCC) report. This damning report on the state of the climate impelled him to think of ways to resist and confront this global crisis. For Bloom, fiction served as an ideal way to raise awareness of climate change issues to reduce the global threat (Brady 2016). In 2012, Bloom produced and promoted a novella by Jim Laughter entitled *Polar City Red.* The novella is set in Fairbanks, Alaska, in the year 2075, when global temperatures in the lower 48 states have become uninhabitable. Although popular attention for *Polar City Red* dried up soon after its release, Bloom called the book an example of cli-fi, or climate fiction, in a press release, and ever since this classification has stuck, defining a new direction in environmental and social literature (Glass 2013).

Climate fiction as a term or sustained genre, form, or mode has yet to receive as much critical attention as one might expect considering the vast amounts of scholarship written about environmental literature over the past 30 years.[1] The following pages of this chapter define and contextualise cli-fi to provide a framework for examining a novel that demonstrates spatial injustices of solastalgia. If climate change is an issue of spatial justice, then cli-fi serves as a mode of interpreting global warming through 'fictional' forms that present both real and imagined spaces. And if climate change is socially produced, constructed through energy systems reliant on polluting fossil fuels, then cli-fi can respond to that social process and reconstruct narratives about the existential threats of climate change. Rather than delineate a prescriptive analysis in this chapter, I want to return to the book's central aim: to scrutinise socially produced spaces, which affects lives, culture, and ways of being and thinking, returning us to the extant environmental crisis. Climate fiction not only draws out explicit concerns about ecological futures (catastrophe or apocalypse), but also challenges the very systems that have produced the climate crisis. Cli-fi is an artistic form representing and resisting climate change and global circumstances of injustice stemming from energy systems and political economies.

Ian McEwan's *Solar* (2010) is one of the most recognised cli-fi novels in climate change criticism, largely because of McEwan's elevated fame as a contemporary literary figure, not necessarily because it offers the most representative example of the genre. As a novel about climate realism, it considers climate change and the social responses and science about it. *Solar* is also a novel partly about alternative energy. It presents a different approach than other cli-fi novels that either imagine dystopian futures resulting in the destruction of Earth or a new Earth discovered elsewhere in the universe (hence the common overlap with science fiction known as sci-fi). *Solar* explores the cultures enmeshed in the climate crisis, but it focuses on everyday lives to stress the absurd reality of the present moment. The ecospatial elements of climate fiction, complementing what has already been termed 'ecohistoricist' (Johns-Putra 2016, 275), draw different perspectives around the everyday lives of people suffering from solastalgia, which is one direct result of climate change. In this chapter, I shall consider how McEwan's *Solar* narrates cli-fi as a form of 'transcendental homelessness' by drawing on George Lukács' *The Theory of the Novel* (1971 [1920]) and extending the interpretation beyond the protagonist's aggravating affectations and into the realm of his fragmented self in the modern environmental crisis. Because of its critical weight in both literary and cli-fi circles, I have devoted an entire chapter to *Solar*. Before exploring the text, the chapter begins by outlining examples of cli-fi and how we might better imagine and understand this textual mode as a reaction to the environmental crisis.

Climate change remains a cultural and political issue as much as a scientific one, which is why writers and specifically novelists have explored what might be the most severe threat to human and nonhuman existence in the twenty-first century. Cli-fi may seem reductive and an unnecessary taxonomy within a corpus of literary genres and styles, which is to say it could be considered 'gimmicky' or a 'fad'. Nevertheless, I want to suggest the name itself invokes a political response to the climate crisis through artistic production, particularly how we might interpret climate change as a social and cultural issue, as well as a spatial one.

One way climate change has affected literary output is through genre (Trexler 2015, 13). But as a collection of work that engages one of the most important social and environmental questions in the twentieth and twenty-first centuries, the relevance of climate fiction should transcend these narrow discussions about how to classify it as a literary form. The greatest tension within debates about cli-fi should be between the existence of catastrophic global warming and the consistently failed attempts to act (Trexler 2015, 9). Cli-fi, perhaps above all else, aims to confront the severe planetary changes. It is a fiction of social justice promoting awareness and action, and like all forms of fiction or artistic production, the quality or aesthetic structures might differ greatly depending upon the author or text.

Because cli-fi is a new form of literature continually transforming and evolving, it remains difficult to classify or define. It does, however, contain some distinct characteristics that often pervade most texts considered to be cli-fi. In a blog entry on the website The Cli-Fi Report, which is still one of the most definitive and timely sources to obtain information about climate change fiction, William A. Liggett explains five features where cli-fi differs from other genres.[2] First, the central theme is explicitly or implicitly about climate change. Second, cli-fi focuses on planet Earth, even though Earth may be compared to other planets or futuristic scenarios of it discussed in the text. Third, it includes information about climate science in the story. Fourth, the story's action illustrates the consequences of changing climates with often catastrophic or sensational results. Fifth, cli-fi synergises the sciences, humanities, and political advocacy into one fictional form (Liggett 2017). Adding to Liggett's useful framing of the genre, cli-fi contains environmental themes separate from climate change or extending beyond it. Cli-fi might also exist in multiple historical times, simultaneously past, present, and future, to show gradations of change, and in various spatial scales in built or non-built environments, or even on other planets, in spaceships, or through virtual spaces. In this regard, the themes in cli-fi challenge time and space.

Cli-fi might include various forms of fiction, incorporating films (narrative), novels, short fiction, drama, and even poetry (Johns-Putra 2016), but for my purposes here I focus on the novel as a literary text because of its ubiquity as a cli-fi form. In *Anthropocene Fictions: The Novel in the Time of Climate Change* (2015), which is the first major book-length study on climate fiction, Adam Trexler specifies the novel compresses many themes into a solidifying whole, particularly through its 'narrative operations'. He suggests that as a literary form, it 'undermines the passivity of place, elevating it to an actor that is itself shaped by world systems. It alters the interactions between characters and introduces entirely new things to fiction' (Trexler 2015, 233). In another recent book about literature of the Anthropocene, *Ecocriticism on the Edge: The Anthropocene as a Threshold Concept* (2015), Timothy Clark comments toward the end of the book in a section focused on climate fiction that Trexler's use of the novel offers a powerful example of what 'fiction can do', which 'is to conceptualise complex, heterogeneous systems', such as national pride, aesthetics, love, social resistance, and flooding, to imagine our environmental futures (179). Finally, in the most recent study, Antonia Mehnert remarks in her book *Climate Change Fictions* (2016) that the cli-fi novel:

> gives insight into the ethical and social ramifications of this unparalleled environmental crisis, reflects on current political conditions that impede action on climate change, explores how risk materializes and affects society, and finally plays an active part in shaping our conception of climate change. (4)

The novel serves as one specific type of fiction that draws upon other traditional 'pulp' forms such as science or romance fiction to reach a wider audience, but through socially expedient topics related to energy, climate, sustainability, mobility, politics, and all of the other potential mechanisms causing or displaying the climate crisis.

In Amitav Ghosh's *The Great Derangement* (2016), which derived from a series of talks he gave at the University of Chicago as part of the Berlin Lectures (2015–2016), he discusses the absence of climate change art and, in particular, how 'climate change casts a much smaller shadow within the landscape of literary fiction than it does even in the public arena' (7). The 'great derangement' for Ghosh is that writers, artists, and filmmakers who produce fiction have shockingly ignored climate change in their creative works. He writes, 'If we believe that the arts are meant to look ahead, open doors, then how is this huge issue of our time, absent from the arts? It's like death, no one wants to talk about it' (Ghosh 2016, 7). Within the arts and humanities, Ghosh focuses even more criticism on literary culture by claiming there have been few literary works about climate change 'taken seriously by serious literary journals' (2016, 7). With respect to Ghosh's point, debating whether fiction writers engage with climate change versus whether those writers produce 'serious' literature are two separate issues.

As a writer focused on social issues throughout his career, Ghosh clearly recognises how literature functions as a symbolic act in society and culture, inviting political interpretation and engagement. He argues that:

> if certain literary forms are unable to negotiate these torrents [global warming], then they will have failed – and their failures will have to be counted as an aspect of the broader imaginative and cultural failure that lies at the heart of the climate crisis. (Ghosh 2016, 8)

Ghosh's latter point underscores the role of literary culture in constructing or reconstructing the climate narrative within society. Cli-fi, whether we debate about its literary value as an art form or not, contains within its very remit the 'broader imaginative and cultural' significance in confronting and ameliorating the climate crisis.

While there has been a dearth of what we now term climate fiction over the past 40 years, compared to overt forms or recognisable environmental fiction, there are now hundreds of representations of cli-fi available. For example, the website Goodreads currently lists 134 novels designated as cli-fi. Trexler's study declares there to be at least 150 examples of cli-fi (2015, 7). Although the debate over how we define or what makes up cli-fi as a genre or style continues, possible global cli-fi writers might include: Margaret Atwood, Paolo Bacigalupi, T.C. Boyle, Octavia Butler, Jonathan Franzen, Rivka Galchen, Maggie Gee, Amitav Ghosh, Emmi Itäranta, Barbara Kingsolver, Ursula K. Le Guin, Arthur Herzog, Doris Lessing, Ian McEwan, Cormac McCarthy, Sarah Moss, Abdul Rahman Munif, Kim Stanley Robinson, Will Self, Jeanette Winterson, and Tim Winton. This selection is largely anglophone, but other global works of cli-fi exist even though they are not as often translated into English nor written about in journals or books. Moreover, some cli-fi works may not attain an elevated literary status, particularly because cli-fi crosses many genres, styles, periods, and boundaries. Traces of climate-attentive fictions exist in the genres of science fiction, comic, thriller, dystopian/utopian, romance, fantasy, horror, speculative, weird, apocalyptic, and teen novels (Trexler 2015, 7). Such literary tropes cross-pollinate with current cli-fi works, which create difficulties in classifying cli-fi as a distinct literary genre, mode, or form. For the sake of focus, this chapter does not address these aesthetic debates about literary forms or genre studies, even though they will likely generate future academic discussions.

Perhaps more than any other common trait of cli-fi is its globality. The universal link we can attribute to all cli-fi is that all cultures experience the effects of climate change in the twenty-first century and have in various ways responded to it. Therefore, cli-fi is not only transnational and planetary; it is also transuniversal, extending out into the cosmos as a way to reflect upon the Earth's relationship with other galaxies. This is partly why so many sci-fi writers have shifted into cli-fi; the themes of planetary survival and the possible need to find other inhabitable planets resonate with current fears for humans on Earth. Cli-fi becomes spatial in the ways it expands scale and scope, but also through how it confronts the socially produced spaces that impel climate action. Cli-fi often deals with the consequences of such actions in the present and future, while also reflecting upon the past as a guide and educational reminder. As a literary style, it draws on the simultaneity of real and imagined spaces to scrutinise our current world and other possible worlds, whether part of Earth or not. John Joseph Adams quips in the introduction of his edited anthology *Loosed Upon the World: The Saga Anthology of Climate Fiction* (2015): 'Welcome to the end of the world, already in progress' (xi). This line captures the spatio-temporal imbrication that climate change, and fiction that focuses on it, must unpack to help us contextualise the ending world around us and force us to arrive at a new one.

If the aim of cli-fi is to confront and address the climate crisis through any form of fiction, then how do we develop a narrative about climate change? How do stories convey a sense of immediacy and mobilise action? How can we normalise these narratives into our daily lives as part of the global conversation about social change? Fiction dealing with climate change must find ways to dramatise aggregate change, to narrate slowness over time, while also highlighting potential dystopian disasters of flooding, food shortages, or intense heat (Macfarlane 2005). Rob Nixon famously calls this effect 'slow violence', which extends to other forms of social violence beyond climate change (2011, 2). The question remains: how might fiction narrate, dramatise, or imagine forms of slow violence both in the moment and over a period of extended (geologic) time? This is the timely and yet complex purpose of cli-fi in the twenty-first century.

Margaret Atwood is one notable pioneer of cli-fi. Her novel *Oryx and Crake* (2003) was among the first recognised cli-fi texts, along with J.G. Ballard's *The Burning World* (1965), Kim Stanley Robinson's *Forty Signs of Rain* (2004), and McEwan's *Solar* (2010), among other writers previously mentioned.

Some consider Arthur Herzog's *Heat* (1977) to be the first novel dealing with anthropogenic climate change (Johns-Putra 2016, 268), which would make sense considering it was published in the same period ExxonMobil found evidence that climate change presents an ecological concern. Despite her forays into cli-fi, Atwood queries if cli-fi might be a productive way of educating younger people about the dangers of climate change. Could this literary mode help them 'think through the problems' and provide solutions? Alternatively, Atwood fears it will 'become just another part of the entertainment business', much like organic has for the food business and sustainability has for the retail industry. More germane to cli-fi as a form of resistance and reconstruction of social spaces through real and imagined art, Atwood acknowledges 'the outbreak of such fictions is in part a response to the transition now taking place – from the consumer values of oil to the stewardship values of renewables' (2015). Perhaps Atwood is right here by indirectly defining cli-fi in the most compelling way so far: as a genre of transition, change, and response that serves as not only a warning, a type of didactic parable, but also as an artistic and creative form aimed at the climate crisis.

Fragmented selves, disintegrating climates

McEwan's *Solar* presents a satirically realist perspective on climate change. It focuses on the remarkably unlikable protagonist, Professor Michael Beard, who is a Nobel Prize winning physicist, and his chaotic life. The novel is divided into three parts, '2000', '2005', and '2009', spanning over a decade. In the first part, '2000', Beard is introduced as a pathetic, middle-aged man whose fifth marriage is falling apart and he is utterly apathetic about his life. After an accident in Beard's house, which kills his postdoctoral researcher Tom Aldous, Beard decides to steal Aldous' research. In particular, Aldous built on Beard's earlier photovoltaic research to think about new ways of producing solar energy sources to combat climate change. This first section of the novel also overviews Beard's failing marriage with Patrice, his work in England's National Centre for Renewable Energy, and an exhibition he takes to the Arctic Circle with a group of climate change artists, the latter of which is of significant focus of my analysis in this chapter. In Part 2, '2005', Beard has a relationship with Melissa, who reveals to Beard that she is pregnant and plans to have their child despite his objection. His career resurges because of Aldous' research (expanded from Beard's earlier photovoltaic work), although Beard fails to acknowledge the ideas belong to the now deceased Aldous. Suddenly, Beard is fired from the Centre because of his misogynist comments that men are essentially better scientists than women, which create a media frenzy. Part 3, '2009', culminates in all of Beard's transgressions when his multimillion-dollar solar project in New Mexico falls apart on the eve of the opening ceremony; Melissa attempts to win Beard back from his new girlfriend; Darlene, a patent lawyer, makes the connection that Beard stole Aldous's research; and a doctor reveals that Beard ironically has what he already suspected was skin cancer on his hand. The novel ends abruptly with Beard's apparent heart attack.

Despite the layers of emotional and psychological drama, *Solar* is comedic, even if overstated and tiresome at the cost of its intended humour. It is also realistic in its depictions of a climate scientist who also serves as an anti-hero within a postmodern environmental epic. As an example of cli-fi, *Solar* stands out not only because it is written by an acclaimed British author, but also because the novel is not set in a post-apocalyptic or dystopian future like so many other cli-fi novels. In this way, *Solar* is a social novel that functions as a unique example of the cli-fi genre. *Solar* offers a slow burn, a cumulative perspective on a protagonist who is a scientist and whose melting life is a metaphor for the world around him. It combines intergenerational relationships, as an hourglass for the future, with climate science. The novel underscores the practical domesticity of the personal through the global enormity of climate justice.

As a way of accepting the urgency of climate change, the novel normalises the scientifically accepted causes and effects of these changes in our lives by integrating Beard's work on alternative solar energy with human dysfunction in relationships and work. From a realist perspective, *Solar* underscores climate is not something removed from daily life. Instead, it is everywhere, affecting everyone in various ways. Besides the satirical and allegorical devices employed, *Solar* raises pressing questions about energy security, air pollution, and peak oil surrounding climate action. Beard may be a central concern and focus, but readers must also look beneath the layers of his self-absorbed and narcissistic behaviour to find everyday questions on the current state of living in a climate crisis. As a result, social reactions to climate change are the novel's other primary focus.

Criticism of *Solar* has been mixed. Regardless, the novel continues to be discussed because of its attention to climate change and because it is written by a well-regarded literary author (Garrard 2009; 2013; Rahmstorf 2010; Shivani 2010; Zemanek 2012; Trexler 2015; Habibi and Karbalaei 2016). The ecocritic Greg Garrard, who initially wrote a gracious review of *Solar* in 2009, later upon reflection acknowledged that the novel 'was considerably worse' than he projected. In particular, the form it took failed, which is described by Garrard as a '*comic allegory* of self-indulgent excess on the part of the protagonist' (2013, 180; original emphasis). The self-indulgence erases any emotional range the protagonist might otherwise achieve. Instead, readers struggle to empathise with Beard and therefore the main theme of the novel: climate change. For Garrard, Beard's oversimplified and obvious tragic fall does not counterbalance the presence of climate science (2013, 180). He argues that the novel 'riles on a cruelly comic analogy between physical weight and carbon emissions that implies both obesity and global warming are failures of self-discipline – a convenient untruth that exonerates the fast food and fast fuel industries' (Garrard 2013, 181). Similar to Garrard, Trexler views Beard's 'selfishness of human character' unlikely to influence or mobilise readers toward 'collective movements' of climate action (2015, 34). Trexler identifies how *Solar* 'suggests that environmentalist efforts to curb human appetites may also disable the creativity needed to address humanity's deepest problems' (2015, 50). Both critiques highlight that Beard's failure as a protagonist undermines possible support for global climate action.

One problem with labelling *Solar* as a cli-fi novel, as opposed to a novel outside of a thematic genre, is that an established canon of what to expect or what may 'work' artistically and socially as an example of cli-fi remains somewhat unclear. The question still remains unanswered because of the novelty of the style: what might a convincing or effective cli-fi novel look like? Returning to Trexler and Garrard's evaluations, does a protagonist carry the burden of producing a successful or unsuccessful work of climate fiction? Is a work of cli-fi only successful if it persuades its audience? Mehnert maintains that the cultural complexity of climate change functions 'within a broader context of discourses and narratives, which implies an awareness of social and cultural spheres through which climatic changes are brought to the fore' (2016, 4). Cli-fi as an artistic mode tackling the topic of climate change must reach a 'broader context of discourses and narratives' that functions in many social spheres. *Solar* uses climate realism, differing in style and tone from most cli-fi novels because of its realistic rather than sensationalised approach. The protagonist's life, as despicable as it may be, similarly highlights the fragmentary effects of climate breakdown in our own lives, not to reduce its value or import, or to ignore its gradual outcome, but to accept it as a part of contemporary existence rather than imagine it as an apocalyptic moment 'out there' in the future and separate from the current crisis.

Despite Beard's expert knowledge of climate science, he distracts himself with tumultuous relationships, excessive eating and drinking, gratuitous sex, and ego-driven professional pursuits. He is a hyperbole of the excess driven North American 'everyman'. Although Beard is British, he arguably acts more like a stereotyped American produced by consumerist cultures of glut and surplus.[3] Life, in other words, is difficult enough to sort out without the added layer of accepting that Earth is facing a climate crisis of epic proportion. Beard acknowledges the reality of climate change: 'But the basic science is in. We either slow down, and then stop, or face an economic and human catastrophe on a grand scale within our grandchildren's lifetime' (McEwan 2010, 149). However, he does not completely accept such the ecological crisis because he struggles to acknowldge his own personal crisis. Beard 'was not wholly sceptical about climate change' and he 'expected governments to meet and take action' (15). He was, nevertheless, 'unimpressed by some of the wild commentary that suggested the world was in peril' (16). He resists fear-based climate narratives verging on the apocalyptic, outright mocking predictions of 'plague-of-boils and deluge-of-frogs' because they do nothing more than create a constant state of 'living at the end of days' that reinforce the 'apocalyptic tendency' to conjure 'yet another beast' (16). The 'end of the world' narrative was 'never pitched in the present, where it could be seen for the fantasy it was', Beard suggests, 'but just around the corner, and when it did not happen, a new issue, a new date, would soon emerge' (16). Paradoxically, Beard's 'heroic' deeds are to resist the stories about catastrophe through an everyday existence of banality.

McEwan's own attempt to discuss the social and ecological issue of climate change echoes earlier calls for action by Robert Macfarlane, one of the earliest UK writers to question the scarcity of climate fiction back in 2005. Macfarlane warns that apocalyptic thinking takes action away from the here and now, which resembles

Beard's resistance to conjuring 'beasts' of fear, by building a sensationalised narrative of doom that if not manifested at certain points leaves environmental thinkers vulnerable to criticism (2005). *Solar* resists the popular urge in other cli-fi novels to rely on or even overstate dystopian futures. Imagining climate futures provides an important analysis but should not do so at the expense of forgetting the future occurs simultaneously in the current moment, or worse, at the expense of oversaturating exaggeration that leads to apathy and inaction in the face of impossibility. *Solar* offers a more balanced approach to climate realism, connecting to readers who have not considered climate science or the contested politics associated with it, as well as those who advocate for climate action. Beard understands the climate science and despite it, or perhaps in spite of it, he cannot resist the human urges driving his own ruin, which, as Garrard maintains, analogises the breakdown of climate action from 'a failure of self-discipline' (2013, 181).

For the rest of the chapter, I examine the spatial elements of the cli-fi novel through the notion of transcendental homelessness, an approach that pushes *Solar*'s interpretation beyond Beard's notable and exaggerated foibles and into the realm of his fragmented self as part of the modern environmental crisis. In *The Theory of the Novel* (1971 [1920]), the Hungarian philosopher Georg Lukács introduces a spatio-temporal literary term called 'transcendental homelessness' as a way to describe 'the urge to be at home everywhere', but which the modern condition disrupts as a 'totality'. He writes that 'the novel form is, like no other, an expression of this transcendental homelessness', wherein 'the ultimate basis of artistic creation has become homeless' (Lukács 1971, 41). Lukács' original intent was to explore modern novels as a response to the artistic homelessness in the Western literary tradition since the Greek epic, a literary form and historical period achieving 'totality'. Many of Lukács' writings outlined the link between social change and aesthetic practice. In *The Theory of the Novel*, Lukács sees historical change as transcendental, marking aesthetic genres responding to social shifts. Although not stated as such, and despite the Edenic idealism underlying Lukács' term as a return to the halcyon days of ancient Greece, these aesthetic genres chart what we might now refer to as socially symbolic acts that mirror the paradoxical relationship between artistic form and life. Lukács argues that the 'problems of the novel form are here the mirror-image of a world gone out of joint. This is why the "prose" of life is here only a symptom' (1971, 17).

While the term may appear primarily temporal, transcendental homelessness has a spatial dimension related to the proximity of home, or sense of place (as a place-home). Lukács' imagined concept of homelessness arose out of the real sociopolitical circumstances of Europe during World War I. His book was a response and confrontation to the fall of 'social-democratic' parties (Lukács 1971, 11). Much like *Solar*'s realistic and satirical response to climate change, Lukács was responding to 'a mood of permanent despair over the state of the world' through the form of the novel and its inherent socially emblematic action. However, Lukács' perspective remains more utopian and idealistic, whereas McEwan's approach does not (Lukács 1971, 12).[4] Many modern European writers and intellectuals felt alienated from their own cultural milieu, or sense of place, because of implicit or specific forms of exile imposed upon them (Neubauer 2009, 6–7). Whether an ideological or physical exile, it created both feelings of liberation and loss.

In his book *Spatiality*, Robert T. Tally Jr. argues that 'this fundamental sense of placelessness' and confusion about place emerging from the modern period created pervasive feelings of anxiety for critics such as Lukács, among other existential writers and intellectuals such as Jean-Paul Sartre and Samuel Beckett, whose work attempted to find meaning in an otherwise absurd existence, which eventually extended into the postmodern condition of a post-industrial society (Tally 2013, 64). The modern novel is itself a form of literary cartography that maps fragmented existence through a narrative form usually focusing on the interior existence of a central character. Transcendental homelessness explains how an individual or collective looks to find meaning in an increasingly unintelligible and dislocated existence where any sense of place of 'home' has evaporated (Tally 2013, 47). Lukács proposes the novel is 'the representative art-form of our age: because the structural categories of the novel constitutively coincide with the world as it is today' (1971, 93). The novel aims to achieve a totality, or wholeness, that cannot be attained in modernity. This presents a paradox because the novel cannot transcend, nor can its subject in the social sphere, and thus it remains without a home, without a total whole.

Lukács captures this modern placelessness, which arguably becomes a postmodern non-place (see Chapter 3 on Augé 2008), through the narrative of the novel in contrast to a romanticised ancient Greece. For Lukács, the modern novel serves as a tool to describe this sense of homelessness in contrast to the epic form, which, for Homer, embodied a transcendental shelter (Lukács 1971, 41). The epic form mirrored the totality of Greek culture, a totality that has become fragmented throughout modernity. However, *Solar* is a novel and not a traditional epic in terms of length, action, literary form, or its portrayal of the central hero, despite some of the novel's epic qualities focused on an anti-hero's postmodern journey to unsuccessfully save the world and himself. Though Beard might wish to be a hero, he functions as an anti-hero posing as a contemporary 'everyman', who is difficult to sympathise with because of his debasingly human actions. Nevertheless, Beard displays a 'fundamental sense of placelessness' within his contemporary existence in a disintegrating social and ecological world. Despite his attempts to use artificial photosynthesis to solve the energy crisis, Beard moves aimlessly throughout the novel, guided by his own uncontrollable corporeality. As a protagonist of climate fiction, his sense of place has evaporated.

What I want to exhibit in this chapter is the spatial aspect of transcendental homelessness as another form of exile in the midst of a contemporary fragmented existence because of climate change. Although framed as a 'historico-philosophical' approach (Lukács 1971, 40), the spatial displacement resulting from transcendental homelessness both in the novel form and mirrored in society underlines a real and imagined form of exile. The fragmented existence, or lack of totality, alienates individuals, which the modern and contemporary novel encapsulates. Individuals are out of place, exiled or homeless, but they must also stay in the same physical geography. This is why the novel form often follows a biographical trajectory to create a type of totality through the protagonist's interior journey. Lukács explains how:

> the novel tells of the adventure of interiority; the content of the novel is the story of the soul that goes to find itself, that seeks adventures in order to be proved and tested by them, and, by proving itself, to find its own essence. (1971, 89)

The psychological journey of the novel is not the hero's adventurous voyage of the epic, but rather one that explores the 'menacing abyss between us and our own selves' (Lukács 1971, 34). As an illustration, Lukács points to the example of Miguel Cervantes' *Don Quixote* (the first global novel, published in 1605). In it, a disillusioned Don Quixote struggles to find meaning in the world he still inhabits. He is out of place and disconnected from home as much as he is detached from reality. In order for Cervantes to find a 'counterweight' to the central character's 'abstract realism', he must balance humour with the sublime (Lukács 1971, 107).

Although seemingly far removed from Don Quixote and Spain in the fifteenth century, *Solar*'s biographical protagonist Beard displays much of the same disillusionment and abstraction from reality. Much like Don Quixote, and later the anxious and dislocated characters of post/modernity in twentieth- and twenty-first-century novels, Beard feels alienated because of the cracked totality of his existence. Beard remains for the entire novel deluded about what is real and what is imagined in both his own life and externally. For example, he cannot commit to relationships (five marriages later) nor to a firm position on climate change, both of which indicate Beard's disillusioned state more than the question of them being real.

Beard's unravelling peaks at the end of the novel when his solar project falls apart, which places him in debt for millions of dollars. He is also being sued for stealing Aldous' research and will likely face criminal charges in England for fraud. The novel ends with Beard's likely heart attack, where 'he felt in his heart an unfamiliar swelling sensation' (283), although Beard, somewhat delusional, interprets it as feelings for his daughter, Catriona, and tries to 'pass it off as love' to himself (283). Readers never know the fate of Beard, just as we do not know the fate of the planet.

The novel form is ideal for searching the inner world of a character, the question of generative or failed transformation. Allegorising Beard with the fate of Earth is less convincing than seeing Beard as a parallel of the fragmented state of society, swirling in lawsuits for his reckless behaviour, without foresight of future progress, and valorising financial opportunism as opposed to social sustainability. Beard is a product of the failing socially produced system in which he lives and works. It is difficult to empathise with Beard, particularly realising he symbolises a cracking global society that has largely ignored climate action. To understand Beard is to accept his inability (and our own as modern subjects) to transcend an increasing disconnection to place and therefore to ourselves. Both theme and form (i.e. *broken* into three separate sections) of the novel mirror Beard's destruction.

The main section of the novel that exemplifies transcendental homelessness is when Beard travels to the Arctic Circle to the Norwegian island of Spitsbergen with a group of artists to see the results of climate change first-hand. The artists on the trip also wanted to create art as a response to the climate crisis in the Arctic. This scene is autobiographical for McEwan, who similarly travelled to the Arctic in 2005. The environmental group Cape Farewell invited McEwan to join other artists and scientists to visit Svalbard, which is a group of Norwegian islands in the Arctic Ocean. It was after this trip that McEwan realised climate change 'just seemed so huge and so distorted by facts and figures and graphs and sciences and then virtue' that he 'couldn't quite see how a novel would work without falling flat with moral intent' (Brown 2010). Regardless, it was this exact trip, after experiencing the disarray of the boot room on the Cape Farewell exhibition, that prompted McEwan to begin writing his first climate change novel. He recalls the disarray of helmets, boots, and gloves needed to enter the Arctic temperatures resembled 'our capacity for rational thinking and gathering data and evidence on the one hand, and on the other these little worms of self-interest, laziness and innate chaos'. This is where Beard's character emerged. McEwan recognised, 'There's something comic about idealism' (Brown 2010).

The ongoing boot room saga during the week-long Arctic trip in *Solar* allegorises the world's inability to prevent catastrophic climate change because of disorganisation and division of basic social systems. Starting on the second day, 'the disorder in the boot room was noticeable, even to Beard. He suspected that he never wore the same boots on consecutive days' (75). Balaclavas, goggles, and snowmobile suits all occupy places on the floor, not their corresponding numbered pegs on the wall. At the midweek mark, 'four helmets were missing, along with three of the heavy snowmobile suits and many smaller items'. The boot room in its disarray, 'the gathering entropy, was the subject of evening announcements' (79). Beard refuses to take responsibility, even though he initially mistook his original peg number 28 for 18 and then resigned himself to take whatever gear he wanted out of frustration. In fact, Beard's initial oversight of the correct peg number creates a ripple effect of chaos whereby no one could be bothered to keep the gear in order. Beard makes the obvious parallel to climate action: 'finite resources, equally shared, in the golden age of not so long ago. Now it was ruin' (79). This comment also reflects the relationship between the totality of Greece, literally 'the golden age', illustrated through the epic form and the fragmentation of modern society, 'gathering entropy', reflected in the novel form.

As others have pointed out, the environmental allegory of the novel parallels Beard's inability to confront his own deficiencies, let alone address global climate disaster (Garrard 2013; Trexler 2015). The chaos that ensues from misplaced gear, disagreements, and unfocused expeditions during a week in the Arctic satirises the larger global community of environmentalists trying to create change. *Solar* does not hide this when Beard muses, 'how were they to save the earth – assuming it needed saving, which he doubted – when it was so much larger than the boot room?' (80). Some might interpret the Arctic trip in this way, but I want to suggest this part of the novel (at the end of Part 1) offers the dénouement rather than serving as a passing scene because it confronts the relationship between art and science to which Beard cynically responds. The boot room episodes appear as minor parts of the group trip to the Arctic Circle. Here, we see what happens when a minor scene develops into one of the more convincing series of events in the novel that reinforces Beard's displacement.

The Arctic trip emphasises how the arts remain fundamental to establishing a sense of place for Beard, brief as the trip may be in the novel, outside of his normal sense of homelessness. Beard begins the entire Arctic trip uncomfortable and disorganised; he regrets accepting the invitation. After a delayed flight from Trondheim to Longyearbyen, Norway, he finally arrives at his lodging. Rather than go to sleep early and wake up alert for the next morning's journey on snowmobiles to the ship, at night 'he lay in bed' and 'ate all the salted snacks in the mini bar, and then all the sugary snacks'. After eventually passing out, 'he was awoken by reception' the next morning because everyone was waiting for him in the lobby (53). Upon waking, Beard realises he is so parched from the medley of salt and sugar snacks the night before that he impulsively imbibes water from the basin, which is ice cold, but 'he drank so deeply, that he developed shooting pains in his face and temples' (53). At this point, the group had been waiting too long, and so they finally left and let Jan, 'a great elk of a man' (53), guide him alone on the 90-minute snowmobile ride to the ship.

This whole episode, although obviously meant to be funny, showcases Beard's painful incompetence and lethargy. The scene reflects his entire existence, disconnected from the world through his personal relationships, professional life, and his body. When hurrying to catch up with Jan on his snowmobile, Beard recalls, 'the long-forgotten experience from his school days, not only being late but feeling ignorant and incompetent and wretched, with everyone else mysteriously in the know, as though in league against him' (54). His present and past modes of being in the world reflect the discord of a modern existence; without a sense of totality with himself or his social community, he is perpetually disjointed. He self-reflects, 'Fatso Beard, always last, useless at team games' (54). This memory causes further 'clumsiness and indecision' in the moment, which triggers a chain of events on the way to the ship that appear comical but perform as pathetic and indicative of his fragmented self.

When snowmobiling at the speed of 40 km to catch up with his team across 115 km of snow and ice, Beard suddenly has to relieve himself of all the ice water he drank just an hour before. Although a Nobel Prize winning scientist, Beard rarely exhibits intelligence or foresight. Here, he attempts to urinate outside in sub-zero temperatures, not recognising this would be physically harmful. Once he finishes, 'he discovered that his penis had attached itself to the zip of his snow-mobile suit' and once 'he pulled tentatively' to dislodge the attachment, 'he experienced intense pain' (60). As a remedy, he quickly pours Brandy from his flask on the zip to free his member, but he is uncertain of the damage because he cannot feel anything in the extreme cold. Jan loses patience and demands Beard get on the back of the snowmobile with him, at which time Beard believes that part of his penis has frozen off: 'Something cold and hard had dropped from Beard's groin and fallen down inside the leg of his long johns' (61). He later realises once on the boat it was only his frozen container of lip salve.

This episodic adventure of everyday experiences of sleeping, eating, and travelling, which are not arduous heroic tasks, epitomise Beard's disjointed existence in the world. Beard's perceived castration in the urination scene illustrates his inability to create or produce significant change, despite the power he holds in society. The overall picture in a 12-hour period condenses Beard's existence into a series of digestible moments that mirrors many others throughout the novel. Disconnected from his community, his relationships, and himself, he cannot organise his affairs or person (without help) to get on the Nordic ship for the climate change expedition, of which he is the celebrity, without drama and chaos. Similar to the boot room, this scene allegorises humans' inability to manage climate because we cannot manage ourselves.

Although travelling to the ship became an ironic version of the heroic journey, his time with the artists serves as one of the few moments of balance, where he restores a sense of place in an otherwise chaotic existence. The people on the Arctic expedition are all artists and writers interested in documenting and responding to climate change in their art. As it turns out, 'Beard was the only scientist among a committed band of artists' (62). Beard initially scoffs at the idea that art has any value to climate change because he cared 'little for art or climate change, and even less for art about climate change' (74). Regardless, he mostly 'listened' to what he considered 'idealists in such concentration', which made him simultaneously 'intrigued, embarrassed, constrained' (75). After repeatedly being asked to assert his theories as a famed climate scientist, Beard finally offers his thoughts, though he initially takes a privileged position and uses a condescending tone when talking to the group of artists. After all, he 'was among scientific illiterates and could have said anything' (76). One artist, Stella Polkinghorne, reminded the group that 'Beard was the only one here doing something "real," at which the whole room warmed to him and applauded loudly' (76). Notwithstanding being artists, 'he was touched' by this acknowledgement, despite not usually caring 'much for what others thought' (76).

This mood does not last long, however, and Beard eventually could not resist responding. After 'just finishing his eighth glass of wine', he lauded his scientific knowledge over the artists. A novelist named Meredith speculated about a correlation between Heisenberg's uncertainty principle, 'which stipulated that the more one knew of a particle's position, the less one knew of its velocity, and vice versa encapsulated' the global issue with morality (77). After pounding his fist on the counter, he indigently responds, 'So come on. Tell me. Let's hear you apply Heisenberg to ethics. Right plus wrong over the square root of two. What the hell does it mean? Nothing!' (106–7). Beard's conclusion is final; climate science and climate ethics are not connected.

Beard's response to the artists' attempts to meet him in his area of expertise was brash. He later reflects upon his own feelings of inadequacy around so many people who help the world, create awareness over a pressing issue that although Beard is an expert on, he either uses only for financial opportunity or feeding his professional ego. Beard may not be the only one, if he even is, doing anything 'real' to save the world from climate disaster. Beard reveals that it would not have been

> possible that he would be in a room drinking with so many seized by the same particular assumption, that it was in its highest forms – poetry, sculpture, dance, abstract music, conceptual art – that would lift climate change as a subject, gild it, palpate it, reveal all the horror and lost beauty and awesome threat and inspire the public to take thought, take action, or demand it of others. He sat in silent wonder. (79)

This moment reduces Beard's scientific superiority to a point he can understand the relevance of art to climate science or other relevant ecological issues. Addressing climate change through artistic production draws on these symbolic acts as much as climate science explores it through equations and experiments. After this scene, Beard relaxes and finds connection with the novelists and artists attempting to address climate change in their creative works.

On a trip that would seem unexpectedly disorganised, Beard finds a totality for a brief moment with the artists and establishes a sense of place in the Arctic. He confesses the transcendence of the Arctic trip on his character, even at his own initial reluctance and continued attempts of self-sabotage: 'Everything about this trip had conspired to reduce him' (71). His brief reduction or transformation equates to feeling at home with himself. After almost being attacked by a polar bear, where he fails to escape because he was 'pressing the headlight switch' instead of the starter on the snowmobile, and where Jan had to bail him out, he found more peace and sense of perspective. Not prone to reflection or introspection, Beard thinks 'he was near the top of the world' (73). Although such a thought was geographically induced, reflecting on his cartographic coordinates in the northern hemisphere, it triggers a 'musing that was comforting' (73). He admits that 'his life was about to empty out and that he must begin again, take himself in hand, lose weight, get fit, live in a simple, organised style' (73). Work also comes up for Beard on his northern sojourn as something he wanted 'to be serious at last about' (73).

By the end of the trip, 'Beard found himself in a mellow state – an unfamiliar cast of mind for the morning'. Unusually, he 'was not even hung over' (80). Present to himself and to his surroundings, which he could not be when arriving a week earlier because he was asleep on the back of Jan's snowmobile after his penis incident, Beard finally experiences the grandeur of the Arctic. He notices 'the frozen shores of the fjord' with 'deep cuts, trenches, in the ice' (80). In perhaps the only moment of self-reflective lucidity outside of his own narcissism, at the end of the trip after connecting with the community of artists he 'was suffused with the pleasant illusion of liking people' and feeling 'unusually warm toward humankind' (81). He 'was still smiling' when walking to his 'twin-propeller plane' returning to England, 'to the mess he had almost managed to forget' (82).

Except for this one lucid week in the Arctic, which briefly transforms Beard through the interaction with climate artists, the novel portrays a dislocated and fragmented society mirrored by an entirely disconnected protagonist. Beard's own homelessness as a modern anti-hero captures the state of a contemporary culture in the midst of an existential climate and ecological crisis. At one point in the novel, he claims modern existence arose for two reasons. First, there were the 'very clever monkeys' that signified evolutionary thinking rooted in Darwinism, and which ultimately leads to contemporary forms of scientific analysis. Second, and almost in tandem, is 'cheap, accessible energy' (148). As discussed in Part II, 'Oil', of this book, cheap forms of energy from fossil fuels of coal, oil, and gas have fuelled what Beard indirectly confirms as a petromodernity. Within such an existence, people are not only dislocated and exiled, but they are also facing climate breakdown because of warming temperatures, flooding, and droughts. Both evolution and cheap polluting forms of energy contradict one another. This also shows a contrast to Beard's transcendent Arctic experience. Fragmented modern existence creating non-places of homelessness ultimately mirrors Beard's experience of the world and others like him, a world with climate threats resulting from inactive social systems unable to change. Is it possible for Beard or others like him to transcend his own homelessness for longer bursts of time? Will global societies evolve beyond polluting forms of energy and address our environmental futures?

Climate homelessness

In a review of Edan Lepucki's cli-fi novel *California*, Sarah Stone articulates the power of novels in the fight against climate change. She writes that if 'we survive . . . it will be in part because of books . . . which go beyond abstract predictions and statistics to show the moment-by-moment reality of a painful possible future' (Stone 2014). The conclusion of *Solar* points to social systems, which for Beard have little value. Regardless, Beard reflects at the end of the Arctic expedition about how humans remain the issue. 'Science' was 'fine' and 'who knew, art was too, but perhaps self-knowledge was beside the point' (81). The boot rooms of the world need 'good systems' to prevent 'flawed creatures' from misusing them. Good 'laws would save the boot room' with 'citizens who respected the law', rather than leaving the world to 'science or art, or to idealism' (81). In contrast to Beard's reliance on 'the law', one but not the only part of climate action, artistic production yields an undeniable effect. For example, it triggered Beard's momentary transcendence in the Arctic even though he quickly returned to his everyday existence of self-involvement. Beard's only transformation characterised wholly without the protagonist's inability to alter or generate change appears after his week in the Arctic with a group of sculptors, choreographers, writers, poets, and musicians. As discussed in the previous chapter, the Inuit politician Aqqaluk Lynge once remarked, 'What happens in the world happens first in the Arctic' (Gurmen 2013). Beard's fleeting transcendence from his existential homelessness happened in the Arctic among a group of artists who, unlike Beard, respond to and actively generate 'real' change.

The narrative structure of *Solar* offers readers a type of climate realism, a way of understanding the everyday realities of contemporary climate change. The less obvious realism in *Solar* is a form of 'scientific realism', as Trexler points out (2015, 34), which implies that scientific techno-utopian fixes will save us from catastrophe, and a realism that accentuates the spatial relations to place-home or homelessness underlying the novel. Beard's homelessness that he mostly cannot transcend becomes the satire. After all, what is more satirical than a protagonist who is a climate scientist, but who fails at addressing climate change? The novel is realistic because in our current age of environmental crisis, many people feel a sense of disconnection from place, or a pervasive homelessness, even if they have physical homes, because of the physically induced stresses of climate change that generate ecological exile. Beard signifies this experience as an everyday person, although readers might resist relating to him on a personal level. He is universalised and hyperbolic for a reason: to magnify the immensity of climate threats. Trexler has argued that 'Beard's immediate desires continually displace action that would prevent climate change' (2015, 47). While this occurs on one level, this also demonstrates how Beard's (or everyone's) desires result from feeling displaced because of climate change.

As a response to the spatial injustice of anthropogenic climate change, *Solar* offers an alternative approach to the cli-fi novel, one that captures the breakup of modern society seeking to transcend experiences of existential homelessness. The age of the novel, for Lukács, coincides with the fragmentation of a real and imagined coherence of totality (Tally 2013, 63). Despite experiencing a loss of place-home, we remain witnesses to our own disconnection to place and destruction of space in the midst of ecological effects from drastically changing climates. Climate change is a disintegration that creates a homelessness people cannot transcend and the modern and contemporary novel as a literary text encapsulates these effects. Loss of place experienced both through migration and the process of solastalgia, forced to stay in in disconnected places, create widespread loss of place. McEwan's cli-fi novel *Solar* characterises this experience through the absurdity of Beard – the embodiment of the human condition, which is selfish, opportunistic, and hedonistic, but which is also inevitably without a sense of place-home in a world increasingly experiencing the phenomenon of climate breakdown.

Notes

1 The following works published in the last six years are exceptions. Some of them constitute full studies whereas others partly write about climate fiction as an established literary form. See Trexler and Johns-Putra (2011), Morton (2013), Tuhus-Dubrow (2013), Clark (2015), Trexler (2015), Johns-Putra (2016), Mehnert (2016), Siperstein et al. (2016), and Whiteley et al. (2016).

2 See Dan Bloom's website, The Cli-Fi Report, which contains blogs, news, article, and books about the genre. Because of the newness of the genre, many writings about cli-fi exist on blogs (such as Bloom's) or online publications such as *The Atlantic* or *The Guardian*.

3 With this in mind, it is not surprising that American reviewers lambasted *Solar* upon its release (Shivani 2010).

4 Lukács acknowledges in the preface to *The Theory of the Novel* that although his book may not seem as creative as 'the expressionist poet's view', he nevertheless was 'expressing similar feelings about life and reacting to the present in a similar way' (1971, 18). Making a comparison between Lukács' *Theory* and McEwan's *Solar* may also appear incongruent, but they both confront social issues through different prose mediums.

Bibliography

Adams, John Joseph. 2015. 'Introduction'. In *Loosed Upon the World: The Saga Anthology of Climate Fiction*, edited by John Joseph Adams, xi–xii. New York: Saga Press.

Atwood, Margaret. 2015. 'It's Not Climate Change – It's Everything Change'. *Matter*, 27 July. Accessed 1 August 2015. https://medium.com/matter/it-s-not-climate-change-it-s-everything-change-8fd9aa671804.

——. 2003. *Oryx and Crake*. Toronto: McClelland & Stewart.

Augé, Marc. 2008 [1995]. *Non-Places: An Introduction to Supermodernity*. London: Verso.

Ballard, James Graham. 1965. *The Burning World*. New York: Berkley Books.

Brady, Amy. 2016. 'The Man Who Coined "Cli-Fi" Has Some Reading Suggestions for You'. *Chicago Review of Books*, 8 February. Accessed 6 July 2016. https://chireviewofbooks.com/2017/02/08/the-man-who-coined-cli-fi-has-some-reading-suggestions-for-you/.

Brown, Mick. 2010. 'Warming to the Topic of Climate Change'. *The Telegraph*, 11 March. Accessed 2 February 2017. www.telegraph.co.uk/culture/books/7412584/Ian-McEwan-interview-warming-to-the-topic-of-climate-change.html.

Clark, Timothy. 2015. *Ecocriticism on the Edge: The Anthropocene as a Threshold Concept*. London: Bloomsbury.

Garrard, Greg. 2009. 'Ian McEwan's Next Novel and the Future of Ecocriticism'. *Contemporary Literature* 50(4): 695–720.

——. 2013. 'The Unbearable Lightness of Green: Air Travel, Climate Change and Literature'. *Green Letters: Studies in Ecocriticism* 17(2): 175–88.

Ghosh, Amitav. 2016. *The Great Derangement: Climate Change and the Unthinkable*. Chicago, IL: University of Chicago Press.

Glass, Rodge. 2013. 'Global Warning: The Rise of "Cli-Fi"'. *The Guardian*, 31 May. Accessed 26 June 2016. www.theguardian.com/books/2013/may/31/global-warning-rise-cli-fi.

Gurmen, Esra. 2013. 'The Loneliness of the Village at the End of the World'. VICE, 22 May. Accessed 23 September 2015. www.vice.com/en_us/article/village-at-the-end-of-the-world-sarah-gavron-interview.

Habibi, Seyed Javad, and Sara Soleimani Karbalaei. 2016. 'Incredulity Towards Global-Warming Crisis: Ecocriticism in Ian McEwan's Solar'. *The Southeast Asian Journal of English Language Studies* 21(1): 91–102.

Herzog, Arthur. 1977. *Heat*. New York: Simon & Schuster.

Johns-Putra, Adeline. 2016. 'Climate Change in Literature and Literary Studies: From Cli-fi, Climate Change Theatre and Ecopoetry to Ecocriticism and Climate Change Criticism'. *WIREs Clim Change* 7: 266–82.

Laughter, Jim. 2012. *Polar City Red*. Denton, TX: Deadly Niche Press.

Liggett, William A. 2017. 'What Sets Cli-Fi Apart from Other Genres? Here Are the Main Elements'. *The Cli-Fi Report*, 10 January. Accessed 12 January 2017. http://northwardho.blogspot.ca/2017/01/what-sets-cli-fi-apart-from-other.html.

Lukács, Georg. 1971 [1920]. *The Theory of the Novel: A Historico-philosophical Essay on the Forms of Great Epic Literature*. Translated by Anna Bostock. Cambridge, MA: MIT Press.

Macfarlane, Robert. 2005. 'The Burning Question'. *The Guardian*, 24 September. Accessed 3 May 2016. www.theguardian.com/books/2005/sep/24/featuresreviews.guardianreview29.

McEwan, Ian. 2010. *Solar*. New York: Doubleday.

Mehnert, Antonia. 2016. *Climate Change Fictions: Representations of Global Warming in American Literature*. New York: Palgrave Macmillan.

Morton, Timothy. 2013. *Hyperobjects: Philosophy and Ecology after the End of the World*. Minneapolis, MN: University of Minnesota Press.

Neubauer, John. 2009. 'Exile: Home of the Twentieth Century'. In *The Exile and Return of Writers from East-Central Europe: A Compendium*, edited by John Neubauer and Borbáia Zsuzsanna Török, 4–106. Berlin and New York: Walter de Gruyter.

Nixon, Rob. 2011. *Slow Violence and the Environmentalism of the Poor*. Cambridge, MA: Harvard University Press.

Rahmstorf, Stefan. 2010. 'Solar'. *RealClimate: Climate Science from Climate Scientists*, 4 May. Accessed 30 October 2016. www.realclimate.org/index.php/archives/2010/05/solar/.

Robinson, Kim Stanley. 2004. *Forty Signs of Rain*. New York: HarperCollins.

Shivani, Anis. 2010. 'Why American Reviewers Disliked Ian McEwan's Solar: And What That Says about the Cultural Establishment'. *Huffington Post*, 2 October. Accessed 31 March 2017. www.huffingtonpost.com/anis-shivani/american-reviewers-mcewan-solar_b_746830.html.

Siperstein, Stephen, Shane Hall, and Stephanie LeMenager, eds. 2016. *Teaching Climate Change in the Humanities*. New York: Routledge.

Stone, Sarah. 2014. 'California by Edan Lepucki'. *SFGATE*, 8 July. Accessed 20 March 2017. www.sfgate.com/books/article/California-by-Edan-Lepucki-5596861.php.

Tally Jr., Robert T. 2013. *Spatiality*. New York: Routledge.

Trexler, Adam. 2015. *Anthropocene Fictions: The Novel in a Time of Climate Change*. Charlottesville, VA: University of Virginia Press.

Trexler, Adam, and Adeline Johns-Putra. 2011. 'Climate Change in Literature and Literary Criticism'. *WIREs: Climate Change* 2: 185–200.

Tuhus-Dubrow, Rebecca. 2013. 'Cli-Fi: Birth of a Genre'. *Dissent* 60(3): 58–61.

Whiteley, Andrea, Angie Chiang, and Edna Einsiedel. 2016. 'Climate Change Imaginaries? Examining Expectation Narratives in Cli-Fi Novels'. *Bulletin of Science, Technology & Society* 36(1): 28–37.

Zemanek, Evi. 2012. 'A Dirty Hero's Fight for Clean Energy: Satire, Allegory, and Risk Narrative in Ian McEwan's Solar'. *Ecozon@* 3(1): 51–60.

9 Irony of catastrophe

The end is nigh

The end of the world seems so absurd it verges on the comical. It is perpetually imagined, while it remains unknown. The 'end' is as much a conceptual state as it is a reality: cycles begin and end causing life and death. Douglas Adams' famous 'trilogy' of five novels grapples with the cosmic dilemma of 'the end of the world' largely through the humour of satire, irony, and parody. What happens when the Earth and everything living on it no longer exist? Adams' answer is simple: 'Don't panic' (1979, 53). For Adams, a British writer, expanding out into the universe presented a creative way to answer some of the weighty questions back on Earth, notably how humans have allowed the Earth to be systematically destroyed. The first and most prominent book of the series, *The Hitchhiker's Guide to the Galaxy* (1979), is a 'story of that terrible, stupid catastrophe and some of its consequences' of when the Earth was annihilated (Adams 1979, 2). Even though Adams' novel is not the primary focus here, it aptly bookends this chapter, as well as concluding the book, because of its relevant theme of narrating catastrophe through irony.

While Adams' futuristic sci-fi series of novels were, at the time, responding to nuclear proliferation and subsequent threat of world annihilation during a tense Cold War, the novels also consider the mounting environmental questions in the 1970s and 1980s and the potential fate of Earth. Adams' story, much like the current story of the planet amid occurrences of climate disaster, is about a 'terrible, stupid catastrophe', which, as *The Hitchhiker's Guide to the Galaxy* highlights, could have been easily avoided. The irony is that about five minutes after a Vogon Constructor Fleet destroys the Earth to build a hyperspatial express route through its star system, based upon demolition orders that were ostensibly on display in the Earth's planning department in Alpha Centauri for 50 years, the Vogons are informed the hyperspatial route is no longer needed and it was all one big mistake. Slartibartfast, one of the original architects of Earth from the universe's 'planet building' world of Magrathea, lamented, 'Five minutes later and it wouldn't have mattered so much. It was a quite shocking cock-up' (Adams 1979, 163).

The inherent relationship between Adams' universal problem presented 40 years ago and the current fate of the planet is nothing short of farcical, which is to say not much has changed. To put it mildly, planetary destruction caused by anthropogenic climate change or nuclear war is an avoidable 'cock-up'. The irrationality of personal and politicised decisions can decide the fate of billions of humans and nonhumans. And yet, it remains a possibility – one the Earth faces based upon the supremacy of the 'human factor' in an industrialised society highly dependent upon carbon-based forms of energy. Amid this uncertainly, the question remains: how do we conceptualise the end, an issue of enormous scope and magnitude? There is a power in apocalyptic narratives, though they tend to be overstated because they verge on overdramatic language and action, and therefore are often less persuasive.[1] For Adams, as well as Miéville and Barry's short fiction discussed in this chapter, the route crosses through the nuanced comedy of irony as a way to narrativise environmental catastrophe.

Irony serves as a comic device often used in literature to make a point through image and metaphor. The language employed in situational irony assumes that a contrary outcome to what might be expected is both illuminating and amusing.[2] In other words, irony is a complex paradox inverting the expected outcome as a surprise. Irony functions not only as a form of humour, but also as a method of inducing empathy and persuasion.

As discussed when introducing *Ecological Exile* at the beginning, values and perceptions can be transformed by appealing to audiences through figurative imagery. Aristotle argues in his celebrated *Rhetoric* (*c.*350 BCE):

> Of the modes of persuasion furnished by the spoken word there are three kinds. The first kind depends on the personal character of the speaker [*ethos*]; the second on putting the audience into a certain frame of mind [*pathos*]; the third on the proof, or apparent proof, provided by the words of the speech itself [*logos*].

Creating an atmosphere for persuasion (i.e. appealing to *pathos* or emotion) is an important piece in the triad of the argumentative appeals, along with appealing to *logos* (reason) and *ethos* (character). *Pathos* translates in Greek as a 'feeling' or process of 'suffering', though contemporary English colloquially uses it to describe emotion. Originally framed by Aristotle as a tool of oral argument, the relevance of *pathos* continues in rhetorical responses to social issues, whether it is through speech, writing, advertising, or social media. Often achieved through stylistic approaches of humour or storytelling, emotional persuasion applies to literary production. Using irony is another way a writer might use *pathos* to depict emotion or feeling. The effect would persuade an audience by drawing on collective understanding between two opposing viewpoints to produce empathy. Appealing to *pathos* through irony to provoke empathy serves as one persuasive method that writers might use to address climate change.

Besides irony functioning as a persuasive device, it operates as a spatial and environmental tool in the way it recognises the *other* or the *unknown*. Irony supports a democratic process by creating a distance between two different belief systems, while also forging connection – that is, establishing a sense of irony among divisions and discursive practices. Claims about 'truth' are often fragmented. Irony challenges such claims through space and displacement. As Morton observes in his book *Ecology without Nature* (2007), 'Irony involves distancing and displacement, a moving from place to place, or even from homey place into lonely space' (98). Calamity would seem to be anything but funny, particularly as a distancing and displacing occurrence often resulting in a loss of place-home associated with the effects of solastalgia.

By adding stylistic comic devices to stories about catastrophe, which heighten the link between tragedy and time (indicating spatial relations), writers distance readers enough to empathise with the subject of discussion. This creates both conceptual and real spaces. Morton uses the word 'strangers' to represent the 'other' in irony. Irony helps to identify and understand the distanced stranger, while loss of irony threatens this connection. He recognises that if 'irony and movement are not part of environmentalism, strangers are in danger of disappearing, exclusion, ostracism, or worse' (Morton 2007, 100–1). The 'stranger' of environmentalism might constitute many forms, but climate change remains a leading contender. Using irony as a way of persuading others to acknowledge the reality of climate change, however that might be imagined in artistic production, acknowledges the unknown of that possible 'worse' or 'disappearing' leading to some kind of end – perhaps the failure of the imagination.

Because global warming functions on such a massive scale, it forces people to become highly imaginative about catastrophe and the possible end of the world. Humans have spectacular abilities of manufacturing fear about the future through imaginative stories in the moment. This returns us to the opening question of this chapter: how do we deal with the subject of our own annihilation? How do writers narrativise disaster? Whether calamity is conceptual or actual, it remains a pertinent scenario to consider in an age of nuclear proliferation and limited energy resources based upon traditional fossil fuel production, which creates food and water insecurity, ecological damage, and climate crisis.

This chapter contends that the use of irony as a stylistic tool for persuasion about climate change in China Miéville's 'Polynia' (2015) and Kevin Barry's 'Fjord of Killary' (2012) underscores spatial elements that resonate with themes of injustice resulting in solastalgia. Both conceptually and stylistically, the short story form allows writers to interrogate specific themes, whether consciously or not, through direct and poignant motifs as a formal totality. Edgar Allan Poe, one progenitor of the modern short story and horror fiction, famously called this process 'unity of impression' because it resembles stylistic approaches to poetry and allows an audience to experience a story in 'one sitting' (1965, 421). Within this literary form, one predominant thematic is clear: short stories, such as 'Polynia' and 'Fjord of Killary', consistently establish an awareness of place.

Inverted icebergs

This first section examines Miéville's short story 'Polynia', as well as 'Covehithe', in part, both from the collection *Three Moments of an Explosion* (2015).

'Polynia' derides narratives of climate apocalypse by creating a scenario in which the world might have to accept uncanny occurrences of environmental anomalies. Space and scale are used to challenge notions of 'the normal', as opposed to 'the weird', which reflects the overall approach of Miéville's short story collection. *Three Moments* vacillates between horror and banality, fantasy and reality, stranger and familiar, and impossibility and rationality, thus challenging our beliefs and perceptions of real and imagined spaces. Through a veneer of irony that is palpable but not overexposed, 'Polynia' looks at how environmental changes related to weather and climate patterns might affect a reader's imagination in strange and unsettling ways.

Miéville structurally challenges and then imaginatively reconstructs the social production of spaces in much of his fiction by opposing unjust structures and systems in urban environments. With a doctoral degree in Marxist theory and international law from the London School of Economics, his background rooted in critical legal studies and historical materialism foregrounds much of his creative output. Miéville's literary production illustrates the abuses of power, instances of injustice (particularly by governments and corporations), and, now with *Three Moments*, the biophysical transformation of the Earth. Perhaps the title and opening story of the collection 'Three Moments of an Explosion' best summarises these elements of his work in the opening sentence: 'The demolition is sponsored by a burger company' (Miéville 2015, 3). The metaphor foregrounds the entire collection and captures Miéville's irony of social and environmental injustice: we are all being destroyed by corporations selling us useless products we cannot resist buying. Regardless of the problems around mass meat consumption and factory farming, many people still love burgers. The paradoxical absurdity echoes the unexpected, and yet slightly believable, destruction of the Earth for a hyperspatial express route in *The Hitchhiker's Guide to the Galaxy*.

Miéville claims, however, his intent of writing fiction is not avowedly political:

> when I write my novels, I'm not writing them to make political points. I'm writing them because I passionately love monsters and the weird and horror stories and strange situations and surrealism, and what I want to do is communicate that. But, because I come at this with a political perspective, the world that I'm creating is embedded with many of the concerns that I have. . . . I'm trying to say I've invented this world that I think is really cool and I have these really big stories to tell in it and one of the ways that I find to make that interesting is to think about it politically. (Anders 2005)

One key point here is that Miéville confesses he invents worlds, often reflecting reality but separate from it. He admittedly reclaims social spaces in literary narratives that embed political themes. In a review of *Three Moments* by the famed science fiction writer Ursula Le Guin, she acknowledges what Miéville does not outwardly disclose: he is a writer 'committed to Marxist principles of social justice, with an intense sensitivity to contemporary moral and emotional complexities' (2015). While a writer's personal aim should not necessarily influence a reader's interpretation of a literary work, it confirms Miéville's pattern of confronting particular injustices in the world. For my purposes in this chapter, Miéville's story elucidates issues of climate justice and place through the Gothic subgenre of the New Weird.

By combining multi-genres of fantasy, horror, steampunk, and science fiction, as well as his self-stylised 'New Weird' fiction, Miéville's writing problematises rather than supports linear notions of empirical reality, the effect of which undermines the primacy and privilege of anthropocentric perspectives. Much like the title of this section, and theme of his story 'Polynia' discussed here, Miéville's fiction functions as a hovering iceberg – unfathomable, improbable, and yet transparent. His fiction also tests our assumptions about economics, politics, technologies, and realities through mysterious circumstances.

Traditional forms of 'weird fiction' (called 'The Weird') were considered to be supernatural tales of the pulp horror genre that included writings by Edgar Allan Poe, Arthur Machen, William Hope Hodgson, May Sinclair, and H.P. Lovecraft. The extended literary movement know as the 'New Weird' underlying Miéville's fiction – among other contemporary writers, such as Kirsten Bishop, Steph Swainston, and Jeff VanderMeer, as well as progenitors like Angela Carter, Octavia Butler, or even Franz Kafka – provides a realist approach to traditional forms of speculative fiction, fantasy, and horror. Ann and Jeff VanderMeer's introductory volume, *The New Weird* (2008), describes the genre as:

> a type of urban, secondary-world fiction that subverts the romanticized ideas about place found in traditional fantasy largely by choosing realistic, complex real-world models as the jumping off point for creation of settings that may combine elements of both science fiction and fantasy. (xvi)

While it would be a mistake to oversimplify Miéville's multifarious catalogue of fiction, it undeniably connects to Gothic themes of the New Weird in its aim and scope, which highlights the narratives of climate disaster in some of his recent work. Most manifestations of Gothic fiction explore disruptions and fractures in the social sphere, thereby explaining instability by other 'weird' means. But the New Weird, similar to the Gothic as a cultural movement, transcends genre. It resists the subdivisions of literary genre. New Weird might be subsequently considered a literary movement because of its constellation of subgenres, subjects, and literary devices that all undermine empirical reality and question traditional social orthodoxy.

Another story in *Three Moments* entitled 'Covehithe' provides an example of the New Weird related to climate and oil that helps to introduce the main discussion about 'Polynia', while also tying in petrocultural themes overviewed in Part II, 'Oil'. 'Covehithe' explains the phenomenon of what happens to old, sunken oil platforms/rigs. The story is set on the Covehithe cliffs near the Suffolk coast in western England. Dughan and his daughter are searching for *P-36 Petrobras*, which is a sunken oil rig that has recently come to life. Inexplicably transformed into an animated mass of steel and concrete, *P-36 Petrobras* contained pipes dangling from 'its roof-high underside', while it 'wore steel containers, ruins of housing like a bad neighbourhood, old hosts, lift-shafts emptying of black water'. Upon arriving on the beach, 'it hesitated. It licked the air with a house-sized flame' (303).[3]

Dughan, who is part of the UN Platform Event Repulsion Unit (UNPERU), takes his daughter to witness one of many such phenomena happening across the Earth. The first instance occurred at the beginning of the twenty-first century in the North Atlantic near Halifax, Canada. The *Rowan Gorilla I*, which is an oil rig that capsized on its journey from Halifax to the North Sea in 1988, rose out of the deep Atlantic Ocean and 'staggered like a crippled Martian out of the water and onto Canada' (304). Miéville further personifies the monstrous movements of the *Rowan Gorilla I*:

> It shook the coast with its steps. It walked through buildings, swatted trucks then tanks out of its way with ripped cables and pipes that flailed in inefficient deadly motion, like ill-trained snakes, like too-heavy feeding tentacles. It reached with corroded chains, wrenched obstacles from the earth. It dripped seawater, chemicals of industrial ruin and long-hoarded oil. (304–5)

Shortly after the *Rowan Gorilla I*'s actions were reported, other offshore oil platforms emerged from the depths of the seas around the world. These included the *Ocean Express* that previously sank off of the Gulf of Mexico; the *Key Biscayne* off of Australia; the *Sea Quest* originally from the North Sea but later went to Nigeria and renamed *Sedco 135C*; the *Ocean Ranger* off of Newfoundland and Labrador; the *Interocean II* in the North Sea; and the *Ocean Prince* off of Sardinia (even though it originally sank in the North Sea). The story imagines an offshore oil rig uprising, where the inanimate nonhuman structures protest, almost supernaturally, by asserting their own agency through action.

In each of these instances, previously sunken oil platforms rose from the bottoms of seabeds and climbed on shore to inject/drill eggs into the Earth. Once this process was completed, which took about 11 hours, the oil rigs lumbered back into the sea. A year after the *Ocean Ranger*'s visit to the 'still-recovering Newfoundland ground into which it had pushed its drill, the first clutch of newly-hatched oil rigs had unburied themselves'. Emerging at night, these 'newly-hatched oil rigs' shook off the Earth and 'stood quivering on stiffening metal or cement legs'. Then, displaying 'tiny helipads', they wobbled 'for the sea' (309).

P-36 Petrobras, a ghostly manifestation mirroring these other haunting examples, had sunk and seemingly been forgotten, despite the traumas that took place during its operation years before. In 'Covehithe', Dughan and his daughter attempt to witness *P-36 Petrobras* clamouring to shore and pushing its drill into the soft fens off of the Suffolk coast to lay its eggs. Despite the explicit horror and monstrosity, there exists a humorous paradox that people would place themselves in such danger to witness this process, which is similar to the petro-tourism discussed regarding McGrath's *The Cheviot, The Stag, and the Black,*

Black Oil (see Chapter 4) and the petrophilia (love for oil) that modern society as a collective share. Part of the absurdity of catastrophe is that people want to watch their own annihilation as a skewed form of entertainment. Oil dispersed in space and through the nonhuman continues to be a spectacle as much as a cause of planetary ruin.

As Fred Botting has noted, the Gothic monster in literature is also a political figure, confronting and haunting moments of social crisis, much like the behemoth oil rigs in 'Covehithe' (1996, 102). The visual and physical qualities of oil rigs present horror. These offshore platforms realistically have hundreds of cables/tentacles that plunge into the seabed, sucking out oil from the Earth's crust. What we might imagine of the offshore sublime – juxtaposing the beauty and terror of offshore oil and gas production – literally comes to life in the story. Recognising such Gothic qualities, the story also challenges the environmental horror of offshore oil with unimaginable and delayed consequences. This story underscores elements of the nonhuman and posthuman in contemporary culture by considering how human agency exists as only one category of a much larger material world.

Little distinction can be drawn in the tone and language between whether this tale is a metaphor or literal horror, and yet it serves both functions equally well. 'Covehithe' confronts the oil and gas industry by animating the monstrous qualities of offshore oil into palpable images; it portrays oil as an embodied subject as much as a material product. With an extensive history of accidents since the mid-twentieth century – many of which have not only led to human fatalities, loss of habitat for animals and organisms, and water pollution, but also to the destruction of many oil rigs now inactive at the bottom of seabeds – offshore oil remains one of the devastating 'anonymities' of neoliberalism (see Chapter 3). But in 'Covehithe', the oil rigs appear as the return of the repressed, haunting society as physical and symbolic reminders that although they are out of sight, they continue to exist. In this way, the story indirectly represents one of the major symptoms of climate change. 'Polynia' performs a similar approach by displaying the unconscious depths of icebergs in plain sight.

This mix of weird, fantasy, realism, and science fiction in 'Covehithe' represents Miéville's work as a whole because it is unpredictable with circuitous preoccupations of unharnessed social systems. In *Three Moments*, topics cohere around a central theme, even though the theme remains elusive and difficult to articulate in words. The collection specifies the imminent catastrophic 'explosion' of the Earth that awaits. Such an explosion is not literal; it often functions metaphorically, as in the example of 'Covehithe' where the oil rig becomes a monster provoked to devour the people who created it (reminiscent of Frankenstein's monster in Mary Wollstonecraft Shelley's eponymous novel). Such a social 'explosion' serves as a vital reminder to humans: never discount what delayed ecological consequences may result from the human factor. The notion of catastrophe assumes that we are already living in the end, an explosion or series of crises sustained over time and space, which in geological time could be for hundreds or thousands of years. Climate change is the explosion, the catastrophe, or the end, but it functions on geological rather than the human scale.

Similar to 'Covehithe', 'Polynia' serves as an example of Miéville's preoccupation with justice, place, the absurd, and the unexplainable. 'Polynia' is broken into 15 short sections without a cohered plot structure. It is written more as an experience, a glimmer of possibility in an alternate universe. The story is set in the past and explained by an anonymous first person narrator who remembers back when the icebergs first appeared 15 years ago, when he was 11 years old. The narrator admits at the beginning of the story he 'was all frenetic' at the time and 'it's hard to say just what happened when' (6). The story is part young adult fiction (YA) in this way because the primary characters are around age 11; the gang of friends include the narrator, Robbie, Sal, and Ian.

One day, enormous floating icebergs appear over London. When these 'cold masses first started to congeal above London', the narrator remembers, 'they did not show up on radar'. They looked like 'glowing things the size of cathedrals, looming above the skyline' (5). They each contained 'microclimates' where 'bitterly cold wind' would float down 'gusting with wispy snow' (6). While Londoners are initially surprised upon seeing the icebergs, they also seem resigned to accept what might otherwise be an impossibility. People continue to live their lives under the various microclimates created by the icebergs over parts of the city. Although formally entitled 'Mass 1', 'Mass 2', and so forth, they have also been given proper names based upon their silhouettes, such as Mass 3: The Saucepan and Mass 5: Ice Skull. The story explains the qualities of the icebergs and the strange events that surround them from the narrator's point of view. For example, Mass 3 is described as 'an iceberg. No more, no less. Cold, austere, barren. Awesome, of course because since when were icebergs not?' The narrator admits, 'except for the fact that it's levitating above London, it seems no more nor less awesome than its cousins in the sea' (9).

What begins as a phenomenon eventually becomes assimilated into mass consciousness. Floating icebergs are now a part of normal London life. People are bemused by the icebergs, but no one in the story quite understands why they exist or how they suddenly appeared in the sky. The opening section of the story explains that when the 'bergs' first materialised, they 'were dismissed as mirages, hoaxes, advertising gimmicks for a TV show' (5). What might be perceived as a catastrophe and a significant warning of atmospheric change, ironically becomes normalised.

As the second story of *Three Moments*, 'Polynia' establishes the style, tone, pace, and subject of the overall collection, creating a 'unity of impression' amid myriad subgenres. In Miéville's writing, the seemingly simple unity is an illusion, one that conjures the reader into believing improbability is not only possible, but realistic. What I ultimately want to claim about the story is that it addresses the massive scale of climate change through irony by inverting the seemingly invisibility of climate change to the visible, also known as the 'iceberg effect', which might be best illustrated through the title.

Polynia (more commonly spelled 'polynya') is traditionally the name of the condition where open water is surrounded by sea ice. But, more recently, it is used as a geographical term for an unfrozen sea or 'lake' in the middle of ice pack.

As the narrator's friend Ian states in 'Polynia', 'That's a lake with ice all round it' (11). With unpredictable jumps in global temperatures and subsequent melting of polar ice sheets at the poles of the Earth, at what climate scientists describe as alarming rates of decline, polynyas continue to form. These open water pools eventually erode larger ice sheets on glaciers, creating rivers that slice through the depths of the sheets. The severed pieces are icebergs in various sizes that float aimlessly at sea for months until they completely melt into the oceans. Polynyas are another symptom of global warming.

One way that climate scientists gauge the state of climate change is by measuring how quickly polar glaciers recede. Glaciers cover over 10 per cent of the land surface of the Earth and 75 per cent of the Earth's fresh water comes from these glaciers. Out of the 99 per cent total volume of the Arctic glaciers, 12 per cent is made of the Greenland Ice Sheet (GrIS) and 87 per cent comprises the Antarctic Ice Sheet (AIS) (Frezzotti and Orombelli 2014, 60). Ice sheets in Greenland and Antarctica are declining more rapidly than scientists predicted.

In Antarctica, for example, the southern end of the peninsula began shedding ice in 2009 and has continued to lose 60 cubic kilometres of ice each year, which equates to 55 trillion litres of water (enough to fill 350,000 Empire State Buildings in the past five years). This data has been confirmed by satellite elevation measurements and can be linked to warming climates. Prior to 2009, receding ice was documented in the northern Antarctic Peninsula from the ice shelves called Larsen A and B. Some glaciers are shrinking about 4 metres each year. Such an extent of ice loss can change the Earth's gravity field (Wouters 2015; Wouters et al. 2015, 899–903). The situation is worse in Greenland; 60 per cent of mass ice loss since 1991 has caused drastic changes in the surface mass balance (SMB). The continual annual mass loss of the GrIS between 1991 and 2015, without enough snowfall to minimise the mass loss, has become the largest contributor to global mean sea-level rise, much more than (but also in combination with) melting of the AIS (Van den Broeke et al. 2016, 1933).

One of the many concerns related to polar ice disappearing is that more rays of sun directly penetrate and heat ocean water, as opposed to hitting the surface ice of glaciers and bouncing back up to the atmosphere. This scenario produces a double negative that exacerbates the melting: the less sea ice the more oceans warm because they absorb solar energy, and the more ocean temperatures increase the more ice mass declines. Ice sheets only absorb 15 per cent of the sun, whereas oceans absorb over 90 per cent.

Polar ice melting additionally contributes to climate change because it releases larger amounts of fresh cold water into oceans. Fresh cold water is denser than salt water so it sinks more rapidly, which alters and eventually reverses global oceans currents. This outcome caused the last ice age. Once the ocean currents change or even stop, the entire macroclimate of the planet shifts. In the short term, the rise of sea level from polar ice melting directly affects the coastal cities dotted around the Earth, which are home to billions of people. Antarctica's thawing process alone has the potential to generate over a metre of global sea-level rise by 2100 and over 15 metres by 2500. Atmospheric warming from greenhouse gases principally cause these periods of melting. Even if carbon emissions from fossil fuel infrastructures can be reduced or even eliminated, warmer ocean temperatures might take thousands of years to recover (DeConto and Pollar 2009, 591; Tedesco and Monaghan 2009, 1–5).

The point here, beyond the obvious fact that polar ice continues to recede causing multiple environmental concerns, is that icebergs breaking off of larger ice shelves signal that glaciers are melting. Icebergs are a contemporary symbol and symptom of the larger climate problem, but one difficult to imagine for a mass audience, especially in a visual or literary text. Icebergs serve as weird phenomena because people rarely see even the tips of them.

In 'Polynia', the broken pieces of ice soar across the sky instead of through vast oceans (mainly out of public view). People who live in the North Atlantic, off of the coast of Newfoundland and Labrador or Iceland or Greenland, witness large icebergs floating near the coasts each spring and summer. While this might be a normal occurrence, the quantity and size of the icebergs have only increased. The icebergs, both in the story and in actuality, signify a larger issue. Miéville, however, relishes in the creative exploration of the symptom, one he contorts to fictive ends. Miéville's story inverts this distant and unseen reality by placing the icebergs in the sky hovering over one of the largest and iconic cities in the world. The narrator explains the 'crags overhead were close to identical to those that had once floated in the Antarctic' (13).

Global warming would raise less scepticism if icebergs floated through waters by New York or London, let alone overhead in the sky. If climate change remains a difficult concept for people to conceptualise because of its delayed effects over time and enormous spatial scale, then something as drastic as suspended icebergs in the atmosphere would likely punctuate such disbelief. As a writer of the New Weird, Miéville excels at engaging such literary devices that defy material reality, challenging the scope and scale of human and nonhuman experience. Writing in the style of the New Weird provides an ability to expand the range of imagination while also extending the real spaces in which it might function.

The story opens by explaining how the bergs materialised over time, even though many only observed them once fully formed, 'glowing things the size of cathedrals, looming above the skyline' (5). The narrator explains, 'They'd started as wisps, anomalies noticed only by dedicated weather-watchers. Slowly they'd grown, started to glint in the early-winter after-noon. They solidified, their sides becoming more faceted, more opaquely white. They started to shed shadows' (5). Again, the iceberg metaphor related to climate change remains somewhat obvious. Icebergs increasingly appear over time as atmospheric temperatures rise. We are told in the story many climatologists explained the phenomenon, but it was not until the effects became visible that people reacted, signalling a lack of imagination at the onset to address the strange phenomenon. Many acknowledged their existence, but some surmised they were a 'hoax' or 'gimmicks'. Once the bergs were a permanent fixture of the London skyline, people could do little but adapt to this new way of life. Weather patterns became unpredictable and unusual throughout London:

'Sometimes the gusts of cold below the ice were particularly bad, became brutal mini-winters, freezing the air into little storms' (13).

In what is known as the 'iceberg theory' or 'theory of omission', people see only the 'tip' of the story or situation, when underneath three quarters of the story remains. The literary theory derives from Ernest Hemingway's *Death in the Afternoon* (1932), where he describes minimalistic and understated writing. The theory is now used as a general colloquialism called the 'iceberg principle'.[4] Only 10 per cent of an iceberg is visible above the water, with the 'bulk' or 'mass' of it hidden below the surface. One might only access the tip or small amount of information or data when most is hidden. There is always more substance than what is initially on the surface of any situation.

Drawing on this universal iceberg principle that also contains the double entendre of climate change, the story's use of drifting icebergs in the sky inverts the actual scenario of how they float in the oceans. 'Polynia' ironically challenges this notion directly by presenting transparent icebergs. The visible sky and less visible oceans invert in the public eye. Most global citizens rarely see the submerged underwater mass of icebergs. In 'Polynia', however, these icebergs appear in their entirety. No one can deny they exist. The atmosphere becomes an ocean. The story visualises the alleged abstraction of climate patterns by concretising the situation through one of its symptoms – icebergs. What lies beyond the proverbial 'tip of the iceberg' becomes optical and serves as a rather physical metaphor of climate change. The tangible effects of climate change do not provide as many concrete examples in one place in time, particularly for those in the Global North who have been less affected by the consequences of sea-level rising, water shortages, and drought.

One of the solastalgic elements in the story is that people are displaced but continue living in their homes, watching and adapting, but without agency, to the changed environments in which they live. Ian later in his life joins the army in 'the new specialist iceberg unit' (22). There are absurd Earth changes against normal expectation, and it creates new forms of suffering, which the iceberg unit confronts. Not only are microclimates immediately altered in subsections of London, many have died either through exploring the icebergs, receiving a broken shard of ice on a car or house, or by engaging in regular activities. When exploring the 'underneaths' (located below a berg), people did not understand the intensity and potentially fatal temperature changes. On 25 December, Mass 6 quickly shifted and dropped 'over the Serpentine Lido, while the Swimming Club were doing their traditional Christmas Day plunge. The downdraft flash-froze the water and a sixty-two-year-old man died'. The news broadcast then states, with an injection of catastrophic irony, '"He was doing what he loved," the club secretary told the news' (18).

The story signals some of the uncanny changes to the Earth. During a period of many days of rain, some people speculated that rainwater might erode the icebergs, dumping enormous amounts of water on London. Some of the scientists working to understand the hovering iceberg phenomenon suggested that there might be a connection between the first appearance of the icebergs and a 'growth of coral across the facades of Brussels' (12). Various forms of 'brain coral, pillar, coral, and prongs of staghorn coral' appeared around the same time. Each week coral would grow on European parliament buildings and surroundings, while contractors attempted to remove the thick outcropping. The 'fishless reef' simultaneously accumulating in Brussels resembles the unexplainable icebergs, but, as the story suggests, the corresponding 'link was mooted' by media and governments (12). 'Polynia' addresses the coral phenomenon again at the conclusion. What is globally known in the story as the 'Great Brussels Reef' provides another global example of uncanny climate and environmental changes that further buttress the unusual iceberg experience. In Japan, although the situation is not yet public in the story, some electronic companies are struggling to sustain a workforce in the factories; the workspaces are 'unusable because they've filled up with undergrowth from the rain forest' (22).

The London Bergs, Great Brussels Reef, and Japan Rainforest all represent the inversion of the iceberg principle – making the invisibility of ecological deterioration visible to the producers of the situation and affecting daily activities related to industry, government, and commerce in metropolitan centres. The story asks how might we live with the drastic consequences of environmental catastrophe in our daily lives? For many suffering from solastalgia, this is a reality, not a fantasy. 'Polynia' inverts the unseen to the seen, the peripheral consequences of industrialisation to the centre. In this way, the story displays the scale of climate change for all to see and understand.

Climate change justice appears in many forms. Miéville offers responses to the Earth's biophysical changes in short fiction. Miéville's third collection of short stories could be categorised as cli-fi, in addition to sci-fi, spec-fi, or the New Weird, in the ways it addresses the anthropogenic destruction of the Earth, specifically through geologic disruptions, such as hovering icebergs in 'Polynia' or monstrous offshore oil rigs evoking revenge on humans in 'Covehithe'.

Elements of absurdist humour also emerge in 'Polynia'. Miéville captures how humans would respond to such a situation by cancelling flights, deploying the military, sending explorations, and issuing news teams. The story subtly mocks human industry and their attempt to solve the unexplainable through clumsy solutions. In one scene, for instance, the scientists were investigating the reasons behind why the icebergs were melting – and its potential dangers to London – when the narrator notes that 'a small, powerful cross-party group of MPs demanded that the government blow up the icebergs with incendiaries' (12). Invoking the theme of 'explosion' here, Miéville derides the ignorance of inaction through sensationalised solutions as an approach to climate action.

Ecophobia

Ecological disaster narratives have ranged in popular culture from sensationalised Hollywood films like *The Day After Tomorrow* (2004), to sci-fi novels like Margaret Atwood's *Oryx and Crake* (2003) and Kim Stanley Robinson's *New York 2140* (2017), or to realist accounts, such as Gavron and Katznelson's documentary film *Village at the End of the World* (see Chapter 7). Climate change lies at the heart of these narratives, but so does the threat or annihilation of place, whether it is New York City, non-places in the future, a small village in Greenland, or, as this final section of the chapter explores, a small fjord named Killary in the west of Ireland. The looming possibility or actual experience of climate catastrophe remains a significant factor as to why and how people feel most affected living in these places.

One way to examine ecological catastrophe stories is through the 'ecoGothic' – a mode of critical enquiry examining ecological approaches to cultural works where concepts of 'nature' or the 'natural' are investigated through fear and monstrosity, as well as the sublime and the supernatural (Gladwin 2014, 39–40).[5] The notion of 'ecophobia', defined as a psychological understanding of an irrational and unexplainable aversion to or fear of the nonhuman world (Estok 2009, 208), provides another way of considering the ecoGothic, and one that underscores terror narratives of destruction and annihilation. The ecoGothic draws out the entrenched biases leading to ecophobia, where humans irrationally fear their relationship to nonhumans. We can trace apocalyptic fears related to catastrophe back to an emergent industrialised society in the nineteenth century (part of the Anthropocene), when people, perhaps for the first time, could comprehend the human capacity to destroy the planet. These anxieties drew on Malthusian 'doomsday' prophecies about population, food supplies, and the relationship between humans and animals (del Principe 2014, 1). Through various forms of imperial colonisation and exploitation, consciousness about limited space regarding human and environmental impact took shape. Limitation, similar to excess, induces anxiety and fear.

In this section of the chapter, I aim to show how Kevin Barry's short story 'Fjord of Killary', in his award-winning collection *Dark Lies the Island* (2012), offers one such example of ecophobia displayed in contemporary Ireland. In Barry's story, similar to Miéville's stories 'Polynia' and 'Covehithe', irony functions as a persuasive tool to address the loss of place caused by disruptive weather patterns. Barry is for Ireland what Miéville is for England, in that they are both counterculture writers reconstructing home-places, even if they represent fictionalised versions of Galway and London. And they both, as Barry admits of himself, 'go completely fucking nuts on the page' to establish an alternative place through uncanny circumstances (Gatti 2016, 49).

Barry's use of language resembles the orality of story in writing, the effect of which creates what he calls 'an intense experience for the reader'. He explains in an interview with Martha Schuman: 'Intensity can take many forms, and very often in my world it's comic' (2013, 61–2). For Barry, it seems language containing humour is literally rooted in place. Much of Barry's fiction contains strong undercurrents of irrational phobias and the absurd, particularly as they relate to specific locations and abnormal environments. Tom Gatti writes that with Barry's fiction, 'everything starts with place'. In the same article, Gatti quotes Barry discussing the nothern County Galway setting for 'Fjord of Killary' in the west of Ireland. Barry admits that there is:

> the huge fucking presence of this big, black, throbbing ocean, which has an extreme effect on our psychology. And the Weather it's putting across us all the time . . . it's a fundamental part of what makes us who we are. It's an extreme place. (2016, 44)

Gatti's article and interview with Barry elucidates some of what appears in his fiction, which confronts and examines the destruction of place in both a real and imagined Ireland by tracing the comic anxieties surrounding this potential outcome.

In a podcast largely focused on the social and economic injustices linked to Ireland's International Financial Services Centre (IFSC), Sharae Deckard scrutinises the effects of the financial crisis on water, food, and the environment in what she calls 'the neoliberal regime on the horizon'. Deckard, a scholar who examines associations between the capitalist world-system and the ecological crisis, briefly signals out Barry's 'Fjord of Killary' as a story that underscores the 'impact of the ecological crisis', specifically by highlighting 'the peripheries' that 'bear the brunt of ecological degradation and capital flight'. Using Barry's story as one of many literary examples in Ireland, she argues it represents the sites most affected by 'the future impact of climate crisis' – peripheral areas of the country not able to defend their rights. While Deckard does not discuss the story as a symbol for larger global catastrophes, she pinpoints how it addresses the climate crisis in the ways global warming slowly erodes 'the web of life' (2015).

In the following analysis of 'Fjord of Killary', I maintain not only is this story a warning of the current and impending climate crisis, skilfully pointed out by Deckard, but it displays the escalating climate scenario using narratives of catastrophe that invoke ecophobia through Barry's signature language style of comedy (i.e. cynicism and irony). The ecophobia surrounding the outcome is equally important as the literary device capturing fear through irony. By way of introduction, I want to highlight another story that explains what Keith Ridgway in his *Irish Times* review of *Dark Lies the Island* calls 'glimpses of lives, glimpses of character, turning usually on a crisis' that frame Barry's fiction (Ridgway 2012). Even the title of Barry's collection signals the explicit 'dark' or Gothic themes of Ireland's geography, providing an ideal tapestry to explore ecophobia. It also serves as an apt metaphor for the darkness hanging over the global islands on the Earth, the impending ecological catastrophe rising like the seawater in 'Fjord of Killary'.

'The Penguins', published in Barry's first collection *There Are Little Kingdoms* (2006), serves as a useful precursor to later depictions of ecophobia in 'Fjord of Killary'. In 'The Penguins', Barry writes about how a plane travelling across the Atlantic must make an emergency landing in Greenland. Through this harrowing tale, Barry deploys heavy doses of dark irony. The humans in the story experience fear – particularly of death – in contrast to their forced relationship with the nonhuman world, which in this case is the bleak and menacing icefields of Greenland. The added layer of farce in the story exists when the stranded passengers resort to dancing like penguins to stay alive from freezing to death while simultaneously eating aeroplane pretzels. Despite some of the people dying in the tragic event, irony remains present in the story, even in the closing lines: 'Fear climbs in fearless ascent but always it fades . . . Now everybody is greenlighted. Now everybody is bulletproof' (154). Throughout the story, the notion of place permeates the overarching narrative by showing the irrational disconnection, or ecophobia, many people have with the nonhuman world.

'Fjord of Killary' is similar to 'The Penguins' in tone and style, but it amplifies the theme of ecophobia even further. It begins with a sentence apropos of Barry's writing style, concurrently austere, comical, and bizarre: 'So I bought an old hotel on the fjord of Killary' (154). There is a stylistic similarity to the opening line in Miéville's *Three Moments of an Explosion* – 'The demolition is sponsored by a burger company' (3) – in the ways both lines establish a sense of instability and unpredictability, while also injecting the prose with a glib and yet foreboding sense of the future. The narrator Caoimhin explains how the exacting and unpredictable terrain around the hotel/inn in northern County Galway resembles the people who live there.

The traditional coaching inn (established in 1648) ostensibly accommodated the English writer William Thackeray. Caoimhin, a somewhat successful poet, purchased the inn because he suddenly desired a romanticised life away from the city, thus recalling his literary predecessors. He describes how the idealised place, known as 'the west of Ireland' with 'the murmurous ocean', contains 'rocky hills hard-founded in a greenish light' (30). But these topographies are an 'intense purplish tone that was ominous, close-in, biblical' (31), or similar to what Botting has described many Gothic landscapes to be: 'desolate, alienating and full of menace' (1996, 2). According to Caoimhin, the 'Sky is weirdin' up like I do't know fucking what' (31). These ominous signs lead to the story's ecological catastrophe. Eventually, the inn floods up to the second floor, and right before everyone almost drowns the waters finally abate. The story's plot of the impending flood remains straightforward in this way, predictable to the reader from the beginning, but the intention is more than mere sensationalism. Rather, the story emphasises how Caoimhin reacts to the looming catastrophe of place and the dark irony threaded throughout the story.

The disaster narrative begins immediately as the narrator discusses the 'violent' rain that, coupled with the purchase of the hotel, 'would be the death' of him (27). The extreme weather on this night looks like the 'end-of-the-fucking-world stuff out there' (27). Heavy rainfall is typical for this region of Ireland. But the 'hysterical downpour' of this evening, with 'great sheets of water streaming down from Mweelrea', prompted many of the locals to admit that the weather 'had been among the wettest bank holidays ever witnessed' (33).

Interspersed in this story of misfortune, the Belarusians, many of whom worked for Caoimhin at the inn, are alluded to as vampires through their lascivious sexual appetites for each other. Four out of the six had 'love bites on their necks' after they were 'feasting delightedly on each other' in the back rooms (39). Besides the Gothic flourishes that reinforce the dark atmosphere, the overarching narrative speaks to similar apocalyptic motifs of flooding, fires, and earthquakes. At one moment, Caoimhin looks out his window and sees seven sheep in a rowing boat, 'bobbing about in the vicious waters of Killary', and they seemed 'strangely calm' (43). Images of Noah and the ark materialise here – albeit in a hyperbolic way, and refer back to the 'ominous, close-in, biblical' described earlier – to signal other historical catastrophes caused by weather and resulting in the end of the world.

The leitmotif of catastrophe leading to potential apocalypse produces the ecoGothic elements where humans, particularly Caoimhin, have to accept the non-human relationship or shared sense of biological origin of life during the flooding of the inn. At the point of imminent death, Caoimhin acknowledges, 'the view was suddenly clear to me. The world opened out to its grim beyond and I realised that, at forty, one must learn the rigours of acceptance. Capitalise it: Acceptance' (44). Perhaps this is the answer of the Great Acceleration of the Anthropocene – the Great Acceptance? Acceptance here does not refer to inactivity or agreement; rather, it underscores an acknowledgement of human frailty, the fatal consequences of the human project and its catastrophic effects upon not only other humans, but also nonhumans and the Earth. Acceptance allows for that 'pause' of place that Tuan discuses (see Chapter 1), which separates mere spaces from places connected to home, with a sense of connection to a specific topography. In a moment of catastrophe, Caoimhin feels alignment with the material world.

Irony exists because Caoimhin comes to understand the interconnectedness of all things, but does so at a relatively late stage in his life, while witnessing his own looming mortality just moments away. The *pathos* of the moment persuades him to 'learn the rigours of acceptance', even through the comical absurdity of the scenario – that is, a romantic idea about writing in the west of Ireland quickly turns into misfortune. Does catastrophe have to occur in order for humans to change the systems in which they have produced and continually maintain? Acceptance in this way is a change from ignorance to understanding, from passivity to action.

After this event, when Caoimhin survives the flooding, he displays an immediate transformation. Having to 'accept what was put before' Caoimhin, 'a watery grave in Ireland's only natural fjord' (44), likely resulted from ecological imbalances of weather systems and increased water levels from rising global temperatures due to climate change. Caoimhin exhibits the paralysis and powerlessness that many people experience in the face of the climate change paradox – inaction caused by the perception of impossibility. Caoimhin attempts to connect with the nonhuman by purchasing a remote hotel in a region with fewer than 300 people, but what solidifies his connection to the unknown, unfamiliar nonhuman is the ruinous flooding of the inn that almost erases the boundaries between human and nonhuman, human civilisation and environments.

The portrayal of the topography surrounding the inn as supernatural or eerie with rocky cliffs off of the misty fjord and blanket bogland sprawling all around it produces two effects. First, the Gothic horror of an environmental calamity often symbolises a shift from human to material agency. Humans contain the power to reconstruct systems that affect environments, as in anthropogenic causes of climate change, but they do not have ultimate control of the Earth. The difference between influence and ultimate control underscores the material agency of the nonhuman, whether it is animals, environments, or even so-called monsters in

Gothic literature, such as the oil rigs in Miéville's 'Covehithe', the constructed 'monster' in Mary Shelley's *Frankenstein* (1818), or the sea in Barry's 'Fjord of Killary'.

Drawing from 'new materialism' or 'the material turn' in the sciences and humanities, material ecocriticism is a relatively new effort to examine matter as a non-anthropomorphic interaction between text and world in literary studies. The 'more-than-human' materiality of landforms, cities, toxicity, animals, or bodies (both inanimate and animate matter), which are forms of complex ecologies, inform humans about the world they inhabit. In turn, the material phenomena of the world provide maps to interpret various narratives. Matter can be read as text. Material ecocriticism underscores elements of the nonhuman and posthuman, but it considers how human agency exists as only one category of a much larger material world (Barad 2007; Iovino and Oppermann 2014).

Even though the floodwater is a material form, it serves as one of the main characters in the story; it controls the narrative and functions as a nonhuman text to decipher. In this sense, environmental criticism interprets the distinctiveness of matter in two different ways. First, the nonhuman capacities of matter imbued with some form of identity can be described or represented in various narrative works, such as literary, cultural, or visual texts. Second, the narrative power of matter creates meanings and substances that enter and interact with human lives 'into a field of co-emerging interactions' (Iovino and Oppermann 2012, 79).

Despite the climate-induced calamity, Caoimhin survives and so does the inn. Thrust into the nonhuman world, and forced to confront his ecophobia of rural Ireland, the mountain Mweelrea, and the unforgiving seas surrounding the fjord, Caoimhin admits in the final line of the story that the 'gloom of youth had at last lifted' (45). The gloom of flooding disasters, however, continues to increase around the Earth. The 'Fjord of Killary' reinforces these realities in the twenty-first century, especially in places of the Global North, not only resigned to the ongoing climate problems of the Global South. The narrative structure of catastrophe stories highlight the danger of global climate changes, which produce a relatively unknown result in the age of the Anthropocene on Earth, as well as challenge romantically constructed notions of 'untouched wilderness' that are often ascribed to 'nature writing'. Barry draws on the narrative power of matter as a way to interact with the human and nonhuman relationships.

The dark irony in the story is that Caoimhin, mirroring the urban Dublin population who seek relaxing refuge in the west of Ireland, thought he would be safe and cosy due to his own romanticised constructions of this place. However, the floodwaters in 'Fjord of Killary' are real and brutal, and will only increase as global temperatures rise. Part of Caoimhin's 'acceptance' is the realisation he glamourized this life – idyllic writer's get-away by the sea – which is entirely a social construction of imagined place. The real place materialised only after a harrowing experience with this place, one that forced Caoimhin to mature. The 'gloom of youth' has lifted.

What makes this a compelling story related to climate change is that it focuses less explicitly on environmental change and effects of that change than it does on the personal and social relationships that underpin the central catastrophe. One of these personal relationships is the dislocation between human and material agency that Caoimhin must confront. The sense of place, in connection with Caoimhin's initial fear of it as an ecophobia, looms so large in the narrative it consumes the story, much like the rising floodwaters do to the inn. The material world both in the story and in the real world reflect the fears experienced by both the protagonist and the reader. Despite all the warning signs of the heavy rain and the 'unstoppable drift on high westerly from the Atlantic', as well as the minks and otters moving to higher ground throughout the day, everyone except Caoimhin refuses to notice the impending danger. Even the dogs 'howled again in fright-night sequence' as the waters continued to rise (38). Regardless, the 'hysterical downpour, with great sheets of water streaming down from Mweelrea' created what locals described as the 'wettest bank holidays ever witnessed' (33).

Everyone in the inn was trapped by the floodwaters surrounding the fjord: 'the roadway between hotel and harbour wall had in recent moments disappeared'. Because of this event, the 'roads', as the character Bill Knott exclaimed in the story, 'will be unpassable' (38). Their literal entrapment in the inn is caused by their inability to acknowledge the catastrophe looming around them. The microcosm of this story mirrors the larger issue of climate action – despite the obvious signs, action remains slow or non-existent, signalling the ultimate cost of a lack of imagination and foresight. The story subtly warns we will all eventually be buried in water even though action could prevent this outcome.

In the final analysis, however, Caoimhin understands his sense of place by confronting his own ecophobia and the importance of one's environment, whether it is in 'a watery grave in Ireland's only natural fjord, or a return to the city and its greyer intensities' (44). It was Caoimhin, a self-professed 'townie', who continually questioned the danger of the water rising on the seawall and the sheets of rain with visibility 'reduced to four-teen feet' (33). Prior to the water breaching the harbour walls, Caoimhin enquired in an exasperating tone: 'Seriously, lads, we haven't seen a tide that high, surely? Have we?' (35). He noticed that the 'harbour wall was disappearing beneath spilling sheets of water' (37). Like the floating icebergs in 'Polynia', the flood serves as an explicit reminder of extreme weather changes that demonstrate both the importance and impact of humans' place on Earth, which is shared by many other nonhuman organisms. The protagonist Caoimhin presciently recognises the inn at Killary fjord avoided Cromwell's landing in Ireland in 1648 and it would also 'see out this disaster, too' (45). The catastrophe of climate change could also be averted, but 'accepting' it and taking action remains the key element.

The end is the beginning

In *The Hitchhiker's Guide to the Galaxy*, Earth is destroyed just five minutes before it is revealed to be the long-anticipated 'Ultimate Question'. Earth, according to Deep Thought, the most intelligent computer ever designed, is the Ultimate Question to 'the Answer to Life, the Universe, and Everything' (Adams 1979, 199).

Slartibartfast acknowledges the immensity of this occurrence: 'The million years of planning and work gone just like that. Ten million years, Earthman, can you conceive of that kind of time span?' (191). Slartibartfast then quips rather ironically, 'Well, that's bureaucracy for you' (191).

As a response to the Vogon's senseless decision, another 'New Earth' was being built as an identical replica of the original planet so the Ultimate Question might still be revealed. However, the project was eventually abandoned after the Ultimate Realisation: 'if there's any *real* truth, it's that the entire multidimensional infinity of the Universe is almost certainly being run by a bunch of maniacs' (Adams 1979, 200; original emphasis). While the obvious metaphor of rebirth out of catastrophe appears in the novel, it also signals that social constructions of space and society matter. The 'maniacs' contort space to serve selective hierarchical ends. Similar to arguments questioning the so-called legitimacy of climate change, the '*real* truth' is that the climate narrative is being appropriated by a bunch of 'maniacs' willing to sacrifice this 10-million-year Earth project to build 'an arbitrary hyperspatial express route' or whatever equivalent capitalist development project might be implemented as a metaphor.

This prescient poignant scene in Adams' novel would seem as the definitive example of spatial injustice and solastalgia – especially for the only surviving humans from Earth, Arthur and Trillian, who have to experience intergalactic displacement while continually living in the galaxy with their home-place eliminated. *The Hitchhiker's Guide to the Galaxy* parallels human bureaucracies related to survival on a cosmic scale. To this end, an intergalactic bureaucracy run by 'a bunch of maniacs' decide the fate of Earth – unknown to most of the galaxy as the Ultimate Question – by constructing 'alternative factual' social narratives about the fate of the planet.

As Margaret Atwood states in the opening of Chapter 7, climate change is 'everything change', and to generate an overhaul of the social system of late (neoliberal) capitalism there will be forms of catastrophe, even if they do not include the destruction of Earth. The Earth will likely survive humans' acts of malfeasance, whereas humans and many nonhumans might not. The most plausible scenario is what scientists refer to as the 'sixth extinction', which follows the previous five extinctions and where 65 per cent or more species have vanished. Over half of the species on Earth face elimination in mere decades because of the rapid disappearance of the tropical rainforests, and this prediction does not even consider the added effects of climate change due to uncontrollable greenhouse gases (Foster 1999, 144). Moreover, scientists advanced the infamous Doomsday Clock from 3 to 2.5 minutes to midnight after 2016, a year that witnessed global rises in nationalism and xenophobia, increased rhetoric about nuclear proliferation, and a purposeful disregard for scientific evidence, particularly as it pertains to warnings about carbon emissions from fossil fuels and global climate change. These injustices caused mainly by 'maniacs' have produced larger spaces for a greater planetary catastrophe than what is already occurring.

In both Miéville's 'Polynia' and Barry's 'Fjord of Killary', we see two examples of fiction writers, similar to Adams, grappling with the Ultimate Question of the Earth's survival through dark ironic comedy. To imagine the end is as difficult as conceiving the space-time scale of Earth and, ostensibly, as difficult as imagining climate change. Both stories envisage forms of catastrophe that link to changing climates and the association to the biophysical environment bringing the perceptions of the end much closer to reality. The end ultimately circulates back to a new beginning. Climate change is everything change. The tip of the iceberg in Miéville's story or the acceptance of fate in Barry's story both signify a return to the beginning, reconstructing the spatial injustices related to climate change through perception and imagination of possible ends or annihilation.

The real and imagined result of ironising catastrophe in short fiction is at the heart of this chapter, and specifically in how it relates to spatial injustice and climate change. Examining narratives of disaster in literary texts, or even in other visual texts or images, provokes thinking about alternative futures to avoid further catastrophe or envisioning creating new worlds. And, of course, extinction resulting from climate change serves as one of the most devastating possible futures if we collectively fail to transition from carbon to post-carbon societies on a global scale.

Ending the book with a chapter on the irony of catastrophe seems an apt conclusion because it offers an ambiguous result by cultivating the comical and fearful response to the end of the world. Morton argues the idea of 'world' as a socially constructed human concept of modernity has already ended. He maintains that the 'end of the world is correlated with the Anthropocene, its global warming and subsequent drastic climate change, whose precise scope remains uncertain while its reality is verified beyond question' (Morton 2013, 7). As a scholar in the environmental humanities, Morton offers a spatially informed theory about climate change, among other 'end of the world' issues, through the notion of a hyperobject. The end is for humans, also called the posthuman, whereas nonhuman beings 'are responsible for the next moment of human history and thinking' (Morton 2013, 201). Global warming is simultaneously a real and imagined concern that challenges previous notions of how ontology or epistemology must be re-examined in the Anthropocene. This is particularly the case in the ways we experience or know about changing climates of fossil fuel cultures moving into the future. Spatial approaches to injustices such as climate or oil reinforce how we might continue to think about new ways of being that sustains cooperative existence for humans, nonhumans, and objects.

My task here has been to reveal how literary and visual texts have attempted to address and resolve this issue. If the 'world' ended for humans at the beginning of the Anthropocene, with the first whistle of the steam engine in the late eighteenth century, the irony of catastrophe is that world is already over. The end, however, may not be the end for humans and nonhumans, but rather a transition to a new beginning, one that has yet to completely materialise. How, then, do we create or produce a new world? How can the social and spatial systems of power change in structure and to provide more agency for massive change and avoid stress-based ecological disasters in the places humans, nonhumans, or objects exist?

Unjust spatial circumstances make it increasingly difficult for people to shape their own needs, particularly when these forces have exacerbated environmental degradation challenging how people can construct their own geographies. This results in a loss of place-home – an ecological exile inducing conditions of eco-anxiety, or solastalgia, that strain people's living connection to the Earth. In the meantime, literary and visual texts will continually respond to existing productions of unjust spaces in their attempts to reproduce new ones, conceiving of sustainable energy worlds and ecological futures. As *The Hitchhiker's Guide to the Galaxy* informs, the imaginative impulse of artistic production might just lead us to constructing a new Earth and a new way of being on it.

Notes

1 Heise discusses successful and unsuccessful examples of apocalyptic narratives. In one 'successful' example, she analyses David Brin's science fiction novel *Earth* (1990). The novel demonstrates a way to narrate global perspectives about the environmental dangers of the planet, while also representing a wide range of stories from individuals and communities experiencing globally divergent cultural realities. Narrative used in *Earth* includes both elements of disaster and global representation (Heise 2008, 208).

2 Many forms of irony exist, though this chapter relies on the most common form known as situational irony, but other forms exist in the stories discussed in the chapter. Dramatic irony occurs when an audience knows what the characters in a story, play, or film do not see or know; verbal irony exists when a person says one thing but means another (e.g. sarcasm); and cosmic irony relies on the idea that something outside of a person controls their fate and outcomes, whether it is god, destiny, or randomness.

3 'Covehithe' was first published in *The Guardian* as part of the Oil Stories contest (see Miéville 2011).

4 Hemingway writes in *Death in the Afternoon*: 'If a writer of prose knows enough of what he is writing about he may omit things that he knows and the reader, if the writer is writing truly enough, will have a feeling of those things as strongly as though the writer had stated them. The dignity of movement of an ice-berg is due to only one-eighth of it being above water. A writer who omits things because he does not know them only makes hollow places in his writing' (1932, 192).

5 Different approaches exist within the ecoGothic even though it remains limited. Andrew Smith and William Hughes pinpoint an 'ecologically aware Gothic' with 'roots within the Romantic' (2013, 1). David del Principe approaches it by 'taking a non-anthropocentric position to reconsider the role that the environmental, species, and nonhumans play in the construction of monstrosity and fear'. Del Principe's more extensive definition extends to the construction of the Gothic body (nonhuman, transhuman, posthuman, hybrid, etc.) (2014, 1).

Bibliography

Adams, Douglas. 1979. *The Hitchhiker's Guide to the Galaxy*. New York: Pocket Books.

Anders, Lou. 2005. 'China Miéville – An Interview'. *The Believer*, April. Accessed 13 January 2017. www.believermag.com/issues/200504/?read=interview_mieville.

Aristotle. *c.*350 BCE. *Rhetoric*. Edited by John Lendon and translated by W. Rhys Roberts. MIT Internet Classics Archive, Book 1 Part 2. Accessed 19 January 2017. http://classics.mit.edu/Aristotle/rhetoric.1.i.html.

Barad, Karen. 2007. *Meeting the Universe Halfway: Quantum Physics and the Entanglement of Matter and Meaning*. Durham, NC: Duke University Press.

Barry, Kevin. 2006. *There Are Little Kingdoms*. Dublin: The Stinging Fly.

——. 2010. 'Fjord of Killary'. *The New Yorker*, 1 February. Accessed 9 September 2013. www.newyorker.com/magazine/2010/02/01/fjord-of-killary.

——. 2012. *Dark Lies the Island.* London: Jonathan Cape.

Botting, Fred. 1996. *Gothic*. London: Routeldge.

Brin, David. 1990. *Earth*. New York: Bantam Books.

Deckard, Sharae. 2015. UCD*Scholarcast* 51: '"The IFSC as a Way of Organizing Nature": Neoliberal Ecology and Irish Literature'. In *Irish Studies and the Environmental Humanities*, edited by Malcom Sen. Accessed 26 May 2015. www.ucd.ie/scholarcast/scholarcast51.html.

DeConto, Robert M., and David Pollar. 2009. 'Contribution of Antarctica to Past and Future Sea-Level Rise'. *Nature* 531: 591–7.

Del Principe, David. 2014. 'Introduction: The EcoGothic in the Long Nineteenth Century'. *Gothic Studies* 16(1): 1–9.

Estok, Simon. 2009. 'Theorizing in a Space of Ambivalent Openness: Ecocriticism and Ecophobia'. *Interdisciplinary Studies in Literature and the Environment* 16(2): 203–25.

Foster, John Bellamy. 1999. *The Vulnerable Planet: A Short Economic History of the Environment*. New York: Monthly Review Press.

Frezzotti, Massimo, and Giuseppe Orombelli. 2014. 'Glaciers and Ice Sheets: Current Status and Trends'. *Rendiconti Lincei. Scienze Fisiche e Naturali* 25: 59–70.

Gladwin, Derek. 2014. 'The Bog Gothic: Bram Stoker's "Carpet of Death and Ireland's Horrible Beauty"'. *Gothic Studies* 16(1): 39–54.

Gatti, Tom. 2016. 'Laughter in the Dark'. *New Statesman*, 24–30 June: 44–9.

Heise, Ursula K. 2008. *Sense of Place and Sense of Planet: The Environmental Imagination of the Global*. Oxford: University of Oxford Press.

Hemingway, Ernest. 1932. *Death in the Afternoon*. New York: Scribner.

Iovino, Serenella, and Serpil Oppermann. 2012. 'Material Ecocriticism: Materiality, Agency, and Models of Narrativity'. *Ecozon@* 3(1): 75–91.

——, eds. 2014. *Material Ecocriticism*. Bloomington, IN: Indiana University Press.

Le Guin, Ursula K. 2015. '*Three Moments of an Explosion* by China Miéville – Masterfully Horrific SF'. *The Guardian*, 29 July. Accessed 16 January 2017. www.theguardian.com/books/2015/jul/29/three-moments-explosion-china-mieville-masterfully-horrific-sf.

Miéville, China. 2011. 'Covehithe'. *The Guardian*, 22 April. Accessed 20 November 2015. www.theguardian.com/books/2011/apr/22/china-mieville-covehithe-short-story.

——. 2015. *Three Moments of an Explosion*. New York: Del Rey.

Morton, Timothy. 2007. *Ecology without Nature: Rethinking Environmental Aesthetics*. Cambridge, MA: Harvard University Press.

——. 2013. *Hyperobjects: Philosophy and Ecology after the End of the World*. Minneapolis, MN: University of Minnesota Press.

Poe, Edgar Allan. 1965. 'The Philosophy of Composition'. In *Poe: Selected Prose and Poetry*, edited by W. H. Auden, 421. New York: Hold, Rinehart & Winston.

Ridgway, Keith. 2012. 'Fearless, Entertaining and Determined to Have His Way with Cliché'. *The Irish Times*, 31 March. Accessed 23 January 2017. www.irishtimes.com/culture/books/fearless-entertaining-and-determined-to-have-his-way-with-cliche-1.492988.

Schuman, Martha. 2013. 'The Order of Things: PW Talks with Kevin Barry'. *Publishers Weekly*, 1 July: 61–2.

Shelley, Mary. 1818. *Frankenstein; or, the Modern Prometheus*. London: Lackington, Hughes, Harding, Mavor & Jones.

Smith, Andrew, and William Hughes. 2013. 'Introduction: Defining the EcoGothic'. In *EcoGothic*, edited by Andrew Smith and William Hughes, 1–14. Manchester: Manchester University Press.

Tedesco, Marco, and Andrew J. Monaghan 2009. 'An Updated Antarctic Melt Record through 2009 and Its Linkages to High-Latitude and Tropical Climate Variability'. *Geophysical Research Letters* 36(L18502): 1–5.

Van den Broeke, Michiel R., Ellyn M. Enderlin, Ian M. Howat, Peter Kuipers Munneke, Brice P.Y. Noël, Willem Jan van de Berg et al. 2016. 'On the Recent Contribution of the Greenland Ice Sheet to Sea Level Change'. *The Cryosphere* 10: 1933–46.

VanderMeer, Jeff. 2008. 'Introduction: The New Weird: "It's Alice?"'. In *The New Weird*, edited by Ann VanderMeer and Jeff VanderMeer, xvi. San Francisco, CA: Tachyon.

Wouters, Bert. 2015. 'Sudden Melt Strikes Glaciers on the Antarctic Peninsula'. *Arstechnica*, 25 May. Accessed 23 January 2017. http://arstechnica.com/science/2015/05/sudden-melt-strikes-glaciers-on-the-antarctic-peninsula/.

Wouters, Bert, Alba Martin-Español, Veit Helm, Thomas Flament, Melchior van Wessem, Stefan Ligtenberg, et al. 2015. 'Dynamic Thinking of Glaciers on the Southern Antarctic Peninsula'. *Science* 348(6237): 899–903.

Index

Locators in *italics* refer to figures. Mc is filed as Mac.

Helping you to choose the right eBooks for your Library

Add Routledge titles to your library's digital collection today. Taylor and Francis ebooks contains over 50,000 titles in the Humanities, Social Sciences, Behavioural Sciences, Built Environment and Law.

Choose from a range of subject packages or create your own!

Benefits for you

» Free MARC records
» COUNTER-compliant usage statistics
» Flexible purchase and pricing options
» All titles DRM-free.

REQUEST YOUR FREE INSTITUTIONAL TRIAL TODAY

Free Trials Available
We offer free trials to qualifying academic, corporate and government customers.

Benefits for your user

» Off-site, anytime access via Athens or referring URL
» Print or copy pages or chapters
» Full content search
» Bookmark, highlight and annotate text
» Access to thousands of pages of quality research at the click of a button.

eCollections – Choose from over 30 subject eCollections, including:

Archaeology
Architecture
Asian Studies
Business & Management
Classical Studies
Construction
Creative & Media Arts
Criminology & Criminal Justice
Economics
Education
Energy
Engineering
English Language & Linguistics
Environment & Sustainability
Geography
Health Studies
History
Language Learning
Law
Literature
Media & Communication
Middle East Studies
Music
Philosophy
Planning
Politics
Psychology & Mental Health
Religion
Security
Social Work
Sociology
Sport
Theatre & Performance
Tourism, Hospitality & Events

For more information, pricing enquiries or to order a free trial, please contact your local sales team: **www.tandfebooks.com/page/sales**

The home of Routledge books

www.tandfebooks.com

For Product Safety Concerns and Information please contact our EU
representative GPSR@taylorandfrancis.com
Taylor & Francis Verlag GmbH, Kaufingerstraße 24, 80331 München, Germany

www.ingramcontent.com/pod-product-compliance
Ingram Content Group UK Ltd.
Pitfield, Milton Keynes, MK11 3LW, UK
UKHW021613240425
457818UK00018B/530

* 9 7 8 0 3 6 7 2 7 1 1 1 4 *